D0623012

This book provides a complete and clear introduction to the use of adhesives to form load-bearing joints in bridges, civil engineering and building structures.

Recent advances in adhesive technology have led to a rapid growth in the use of adhesives in load-bearing joints in civil engineering applications such as buildings and bridges. In many cases, the use of adhesives can prove more convenient, less expensive, stronger, and more durable than traditional methods of joining. The first part of this book addresses the crucial factors involved in the formation of a successful adhesive joint, including adhesive selection, surface preparation, joint design, fabrication and protection from the environment. The second part of the book describes the growing uses of adhesives to repair and strengthen existing structures, in addition to describing their use in new construction and applications which are now being developed for the future. The connection between the two parts of the book is brought out wherever possible so that the links between theory and practice are emphasized. The book gives the reader all the information required to fully exploit the economic and technical advantages of adhesives.

Professional civil and structural engineers in higher education and industry will find this book an invaluable source of information on a technology of increasing importance. The book will also be useful to graduate students of construction.

Adhesives in civil engineering

Adhesives in civil engineering

G. C. MAYS

Head of Civil Engineering Group,
Cranfield Institute of Technology,
Royal Military College of Science

A. R. HUTCHINSON

Deputy Head of Joining Technology Research Centre,
School of Engineering
Oxford Polytechnic

CAMBRIDGE
UNIVERSITY PRESS

Published by the Press Syndicate of the University of Cambridge
The Pitt Building, Trumpington Street, Cambridge CB2 1RP
40 West 20th Street, New York, NY 10011-4211, USA
10 Stamford Road, Oakleigh, Victoria 3166, Australia

First published 1992

Printed in Great Britain by Redwood Press Limited, Melksham, Wiltshire

A catalogue record of this book is available from the British Library

Library of Congress cataloguing in publication data
Mays, Geoffrey.
Adhesives in civil engineering / G. C. Mays, A. R. Hutchinson.
p. cm.
Includes bibliographical references and index.
ISBN 0-521-32677-X
1. Adhesives.   2. Joints (Engineering)   I. Hutchinson, A. R.
II. Title.
TA455.A34M39 1992
624.1'899--dc20   91-36540 CIP

ISBN 0 521 32677 X hardback

KW

Contents

Contents

Contents

APPENDIX
COMPLIANCE SPECTRUM FOR A TWO-PART COLD-CURE ADHESIVE FOR STRUCTURAL BONDING OF STEEL TO CONCRETE

Preface

The science of adhesion is truly multi-disciplinary, demanding a consideration of concepts from such diverse topics as surface chemistry, polymer chemistry, rheology, stress analysis and fracture mechanics. On the other hand adhesives and bonding form a part of our everyday lives and there are very few of us who have not used 'glues' in one form or another for DIY jobs at home or work. The difficulty arises in translating an understanding of the complexities of adhesion science to some fundamental ground rules for achieving a strong and durable bonded joint which can be relied upon to transmit significant externally applied forces throughout the life of the structure or component. The construction industry, particularly the civil engineering side, presents a unique situation in respect of the conditions under which adhesive joints may have to be formed, the materials to be bonded and the relatively long design life required as compared with other industries such as aerospace and automotive.

The purpose of this book, therefore, is to provide the practising civil engineer with an insight into the factors involved in the formation of a successful adhesive joint, viz. adhesive selection, surface preparation, joint design, fabrication and protection from hostile conditions in service. The application of these principles is then illustrated by reference to examples of current uses in repair, strengthening and new construction, and the potential for future developments. In so doing it will enable the reader to fully exploit the possible economic and technical advantages over more conventional methods of joining, particularly for dissimilar material combinations, the exploitation of new materials and for the development of novel design concepts.

Our thanks are due to the following organisations who have supplied information or photographs as indicated.

Allan H. Williams Ltd
British Alcan Aluminium

Celtite (Selfix) Ltd
Ciba Geigy
Chemical Building Products Ltd
Dow Corning
Elsevier Applied Science Publishers Ltd
European Building Components
Fibreforce Composites Ltd
GEC Reinforced Plastics
General Electric Company
Hilti Ltd
Laminated Wood Ltd
Ove Arup & Partners
Resin Bonded Repairs Ltd

Our thanks are also due to past and present members of the Wolfson Bridge Research Unit at the University of Dundee and its Advisory Committee who initiated many of the concepts for research into potential construction applications for structural adhesive bonding. Particular thanks, however, go to Jan Price who typed the complete manuscript for this book and to Jean Mosley who prepared the artwork.

PART ONE

Adhesives and adhesion

CHAPTER ONE

Introduction

Adhesives can offer substantial economic advantages over more conventional methods of joining. Whilst the building and construction industries represent some of the largest users of adhesive materials, few applications currently involve adhesive joints which are required to sustain large externally-applied forces. However, recent advances in the science and technology of adhesion and adhesives suggest that structural adhesives have enormous potential in future construction applications, particularly where the combination of thick bondlines, ambient temperature curing and the need to unite dissimilar materials with a relatively high strength joint are important. Indeed adhesive bonding, either alone or in combination with other methods of fastening, represents one of the key enabling technologies for the exploitation of new materials and for the development of novel design concepts and structural configurations.

1.1 Definitions and bonding

An adhesive may be defined as a material which, when applied to surfaces, can join them together and resist their separation. Thus adhesive is the general term used for substances capable of holding materials together by surface attachment and includes cement, glue, paste, etc. There is no universally accepted definition of a structural adhesive, but in the following chapters the term will be used to describe monomer compositions which polymerise to give fairly stiff and strong adhesives uniting relatively rigid adherends to form a load-bearing joint.

The term adhesion refers to the attraction between substances whereby when they are brought into contact work must be done in order to separate them. Adhesion is an important phenomenon in science as well as in engineering, but it is used in a different sense. The engineer uses experimentally determined values, which describe joint behaviour under specified conditions, in order to classify the bond or adhesion between two phases. To the physical chemist adhesion is associated with intermolecular forces acting across an interface, and involves a consideration of surface energies and interfacial tensions. The modern science of adhesion is concerned at a fundamental level with increasing our knowledge of the nature of the forces of attraction between substances, determining the magnitude of such forces and their relation to measured joint strengths. For many adhesive/substrate interfaces of practical importance, however, there are still unresolved debates concerning the detailed mechanisms of adhesion and the mechanics of joint rupture.

The materials being joined are often referred to as the adherends or substrates. The properties of the composite made when two adherends are united by adhesive are a function of the bonding, the materials involved and their interaction by stress patterns. It can be appreciated that adhesive bonding technology presents some very unfamiliar concepts to all engineers and, in particular, to civil engineers.

The science of adhesion is truly multi-disciplinary, demanding a consideration of concepts from such topics as surface chemistry, polymer chemistry, rheology, stress analysis and fracture mechanics. It is, nevertheless, important for the technologist to possess a qualitatively correct overall picture of the various factors influencing adhesion and controlling joint performance in order to make rational judgements concerning the selection and use of adhesives.

As with any new technology there are both advantages and disadvantages so that when considering the use of adhesives the merits of the main alternative means of joining (e.g. by welding, bolting, riveting and brazing) should be assessed. The main advantages and limitations of adhesive bonding are given in Table 1.1. The opportunities for increased design flexibility and innovation in design concepts is very real, provided that due consideration is given to balancing the needs of the various materials in a bonded assembly. However, the difficulties of ensuring a good standard of surface pretreatment, particularly to enhance long-term joint durability, are very real, as are the difficulties inherent in verifying the integrity of bonded joints.

Table 1.1. *Advantages and limitations of adhesive bonding*

Advantages	Limitations
Ability to join dissimilar materials	Surface pretreatments normally required, particularly with a view to maximum joint strength and durability
Ability to join thin sheet material efficiently	Fairly long curing times frequently involved
More uniform stress distribution in joints, which imparts enhanced fatigue resistance	Poor resistance to elevated temperature and fire
Weight savings over mechanical fastening	Structural joints require proper design
Smooth external surfaces are obtained	Brittleness of some products, especially at low temperatures
Corrosion between dissimilar metals may be prevented or reduced	Poor creep resistance of flexible products
Glueline acts as a sealing membrane	Poor creep resistance of all products at elevated temperatures
No need for naked flames or high energy input during joint fabrication	Toxicity and flammability problems with some adhesives
Capital and/or labour costs are often reduced	Equipment and jigging costs may be high
	Long-term durability, especially under severe service conditions, is often uncertain

1.2   Structure of the book

Part 1 – Adhesives and adhesion

The first part of this book addresses the important factors involved in the formation of a successful adhesive joint, namely the:

(1)   selection of a suitable adhesive
(2)   adequate preparation of the adherend surface
(3)   appropriate design of the joint
(4)   controlled fabrication of the joint itself

(5) protection of the joint from unacceptably hostile conditions in service, including the provision of fire protection for primary structural bonding.

The development of methods for post-bonding quality assurance might also be added to these factors, since a barrier to the general introduction of structural adhesives into construction is the lack of a reliable method of assessing the quality of bonded joints.

The broad overview of materials and applications, both current and potential, which follows in this chapter leads into a classification and characterisation of adhesive materials in Chapter 2. The various types of (particularly engineering) adhesives are discussed briefly, and it is shown how physical and mechanical characteristics are linkable to chemical compositions. It is apparent that adhesives are in general complex and sophisticated blends of many components, and this background serves to familiarise the reader with an introduction to the chemistry and formulation of adhesives.

In Chapter 3, theoretical aspects of adhesion are reviewed with the object of discussing why adhesives stick, before addressing practical aspects of the surface pretreatment of a number of common construction materials. It is shown that merely establishing interfacial contact between adhesive and adherend is often not sufficient in itself to ensure satisfactory performance. Particular, and sometimes elaborate, pretreatment procedures are found to be necessary for maximising joint durability, and this subject is further elaborated in the following chapter.

Chapter 4 discusses the design and mechanical performance of adhesive joints with particular reference to load-bearing assemblies. The problem of joint design is approached from a consideration of the strains and stresses induced in joints as predicted from stress analysis techniques. Design and testing are natural partners, and in this chapter an extensive review of test procedures is made from which a valuable insight into the adhesive layer behaviour in larger-scale joints may be deduced. Finally the factors influencing joint behaviour and service-life are given.

Chapter 5 looks at the process of joint fabrication, discussing the procedures necessary to ensure a reliable outcome and the methods for testing and quality control of the bonding operation. Emphasis is again given to optimising conditions for maximising potential performance, and some consideration is given to methods of protecting the joint from unacceptably hostile conditions in service.

Part 2 – Applications

The second part of this book is devoted to current and potential applications of adhesive materials in construction. Chapter 6 deals with both the repair and the strengthening of concrete structures, covering applications ranging from non-structural patch repairs and resin overlays to externally bonded steel plate reinforcement. The theme of repair and strengthening is extended to applications involving steel, timber and masonry structures. A number of 'case histories' are reviewed and discussed with reference to the successes and failures, and the results of allied research work are presented. In Chapter 7 a number of applications of adhesives in new construction are described, and specific examples are given. The final chapter, Chapter 8, examines the potential for future developments in adhesive usage.

The link between the first and the second parts of the book is emphasised throughout in an attempt to connect theory with practice, highlighting some of the problems and identifying methods for overcoming them.

1.3 Historical development

Sticking things together is a common enough task, and materials exhibiting adhesive properties have been employed in a sophisticated manner since earliest times. Natural adhesives such as starch, animal glues and plant resins have been used for centuries, and are still used widely today for packaging and for joining wood. Rubber-based adhesives were introduced in the shoe and tyre industries towards the end of the nineteenth century, but the birth of modern structural adhesives is generally dated from the early twentieth century with the introduction of phenol-formaldehyde resins. Mainly as a result of the Second World War, many natural products were not available in the early 1940s and this spurred the further development of synthetic resins. The construction of wooden wartime aircraft was, nevertheless, facilitated by the availability of phenol-, resorcinol- and urea-formaldehyde adhesives, and since then reactive formaldehyde-based adhesives have continued to be used in the manufacture of timber-based building elements such as plywood, chipboard, and laminated timber beams.

Over the past four or five decades the natural adhesives have been improved, and there has been an intense development of synthetic adhesives to meet more technically demanding applications. These synthetic polymers and ancillary products, which include thermoplastic and thermosetting types, have been developed to possess a balance of properties that enables them to adhere readily to other materials, to have an adequate cohesive strength and appropriate mechanical characteristics when cured, to possess good durability, and to meet various application and manufacturing requirements.

Thermoplastic adhesives may be softened by heating and rehardened on cooling, and included in this group are polyvinyl acetates (PVACs). Since the 1950s they have been used extensively as general-purpose adhesives for bonding slightly porous materials, from floor screeds to timber; they are, however, sensitive to wet alkaline service conditions, effectively restricting them to indoor use. Similar adhesives suitable for external situations are based on other polymer dispersions such as styrene butadiene rubbers (SBRs), acrylic polymers, and copolymers of vinyl acetate with other monomers. Cyanocrylates, or 'superglues', also belong to this class of thermoplastic adhesives and are very useful for bonding small parts involving plastics, rubber, metal, glass, and even human tissue.

Thermosetting materials are so called because, when cured, the molecular chains are locked permanently together in a large three-dimensional structure; they may, therefore, be regarded as structural resins. Unlike thermoplastics they do not melt or flow when heated, but become more rubbery and lose strength with increasing temperature. Phenolic resins, and their modifications, belong to this group of adhesives and are numbered among the early structural adhesives used extensively within the aerospace industry for bonding metal parts. Epoxides and polyesters also belong to this group of thermosetting adhesives, and they find widespread use in civil engineering applications. Unsaturated polyesters are often used as binders in glass-reinforced plastics, or as mortars in conjunction with stone and cementitious materials. However, high shrinkage on curing, poor resistance to creep and low tolerance of damp conditions significantly restricts their application. Epoxides, on the other hand, are generally tolerant of many surface and environmental conditions, possess relatively high strength, and shrink very little on curing. There are available a range of epoxy materials which cure at ambient or elevated temperatures, whose mechanical and physical characteristics vary widely. Indeed the general term epoxy may

8

include materials which vary from flexible semi-elastic coatings and sealants to epoxy resin based concretes. Epoxy adhesives are available as single- or two-component materials in liquid, paste or filmic form, which may additionally be 'toughened'.

The increasing use of adhesives in a diversity of demanding situations has given confidence in the successful application of synthetic polymers, and has provided the spur for further fundamental research and the development of improved products. In the future it is possible that acrylates and polyurethanes, and their toughened variants, may challenge the epoxides – particularly as they are perceived to be safer to use and less environmentally harmful. Structural silicone adhesives may also be introduced for certain applications where gap-filling and flexibility are required, but where high strength is relatively unimportant; they also possess the added advantage of very high thermal and environmental stability.

1.4 Engineering applications of adhesives

It is clearly impossible to identify and to document all of the applications of adhesives in engineering assembly and fabrication. Many uses are, anyway, either of a relatively trivial nature or else do not place great demands on the adhesive material. The following sections review some of the major applications of adhesives in several different engineering sectors, in order to put a number of the general design and process considerations discussed later in the book into perspective.

Aerospace

It has often been observed that the application of adhesives to metal fabrication, in common with many other technological innovations, was pioneered by the aircraft industry. It is ironic that this industry, in which safety and reliability command paramount attention, should lead the departure from traditional methods of joining. Today adhesives are used to bond critical parts in commercial and military aircraft and helicopters, spacecraft, rockets, missiles and the US Space Shuttle. The American Primary Adhesively Bonded Structure Technology (PABST) Programme, which ran from 1976–81, was an imaginative attempt to advance significantly the use of bonded structures in aircraft. The project involved the construction and testing of an entire adhesively-bonded fuselage section of a military aircraft.

The earliest structural adhesive application was made during the First World War for bonding the wooden frames of Mosquito aircraft; strength was adequate but, by today's standards, moisture resistance was poor. Structural adhesive bonding of metal parts began in the early 1940s with the introduction of Ciba Geigy's Redux 775, phenol formaldehyde-polyvinyl formvar (the trade name is a composite of *Re*search and *Dux*ford airfield), to bond metal honeycomb to metal skins. The de Haviland Comet jet airliner was one of the first civil aircraft to make significant use of structural bonding with Redux 775, followed by the Trident, Nimrod and VC10, and this adhesive is still in use today on the British Aerospace European Airbus; indeed, no other adhesive has such a good and well-proven track record. However, a cure temperature of 150 °C and pressures between 0.2 and 0.7 MNm^{-2} (25 and 100 psi) are required, and this spurred the development and use of alternative products. For example, epoxy-based filmic adhesives are commonly used which require lower curing temperatures and pressures, making the use of autoclaves and the cure cycle less expensive than with the phenolics.

The motivation in the aerospace industry to replace mechanical fasteners with adhesives stems from the desire to prolong aircraft life and to reduce costly maintenance. Rivet holes, for example, are points of weakness where fatigue cracks can form, and metal fasteners can corrode or loosen. Equally important is that aircraft are now designed to include a large amount of composite materials, and the fabrication of honeycomb sandwich panels frequently involves connecting dissimilar materials for the skin and core. The Fokker Friendship F27 airliner employed large amounts of adhesive in both load-bearing and secondary structures, the wing assemblies being tested up to 14.5 million cycles of reverse loading (Fig. 1.1).

The use of material combinations has continued in aircraft such as the Boeing 747, McDonnell Douglas DC10, Lockheed Tristar and Concorde, as designers recognised the high stiffness to weight ratio that can be achieved with these components. The sheet metal skin materials are predominantly aluminium alloys, although titanium and stainless steels are also used for special purposes. The honeycomb material is often aluminium foil but paper, laminated nylon paper and phenolic impregnated glass fibre are alternatives. As much as

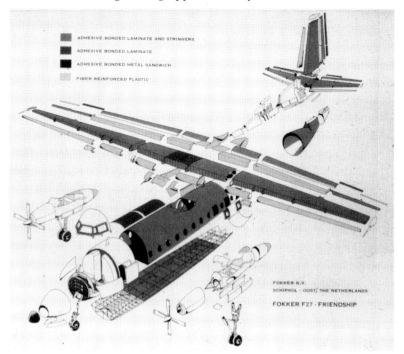

Fig. 1.1. Adhesively bonded components in the Fokker Friendship F27 airliner.

50% of the airframe of modern military aircraft may be carbon fibre reinforced plastic (cfrp) composite, with adhesives being used for primary structural bonding. It is worth noting also that helicopter rotor blades are complex structures which are highly stressed and of limited fatigue life, and are wholly dependent upon adhesive to join the extruded aluminium spar, nomex core, grp skin and aluminium trailing edge.

Honeycomb structures are susceptible to physical damage because of their location and their weakness to loads, applied normal to the skin forces. Airlines and Air Forces are, therefore, developing repair techniques which they can carry out themselves in preference to purchasing expensive replacement components. For *in situ* repairs alternative techniques in terms of surface preparation, adhesive type and curing regime have to be used to those employed during initial manufacture. Surface preparation is vitally important bearing in mind that a typical airliner operating temperature range is from −80 °C to +80 °C, and for many aircraft salt spray conditions may be severe. For the same reason the choice of adhesive is critical. Cold-cure products may be the only alternative for repair work despite their performance disadvantages. Finally, use of the correct curing temperature and pressure is important.

Building

The majority of the adhesive used in the construction industry is concerned with fastening decorative finishing materials to the insides of buildings. For instance, the attachment of ceramic tiles and mosaics to floors and walls, wooden and flexible floor coverings, ceiling tiles, thermal insulation materials, wall veneers, covings, nosings, and so on, accounts for the usage of large amounts of a variety of adhesive materials. A small proportion of adhesive is used in external situations, or for applications in which the structural demands of the adhesive are somewhat greater. Some examples include:

(1) flexible and other roof coverings
(2) resinous grouts for anchoring bolts, ties, service conduits, etc.
(3) joining and attaching internal building panels and elements
(4) structural sandwich panels and cladding panels
(5) fabrication of anodized aluminium window frames
(6) consolidating or joining timber members (e.g. glued laminated timber – 'glulam') (Fig. 1.2)
(7) steel/wood ('wirewood') assemblies for floor beams and roof trusses
(8) structural silicone glazing of glass and cladding materials into curtain walling (Figs. 1.3, 1.4 and 1.5)
(9) attachment of brick slips to concrete
(10) joints between precast concrete units (e.g. Coventry Cathedral, Sydney Opera House roof)
(11) fabrication of special shapes of concrete or clay brick units, such as roof tiles and building blocks
(12) attachment of thin or coated sheet metals and plastics
(13) structural aluminium connections and the compounding of individual extrusions by bonding to create deep structural sections.
(14) strengthening of concrete-framed buildings with external steel plate reinforcement.

Fig. 1.2. Use of glulam structural elements.

The concepts and considerations involved with some of these and other applications were addressed during the period 1988–91 by a European consortium, coordinated by Oxford Polytechnic, within a project entitled Adhesive Bonding Technology for Building Construction.

Weight savings and structural efficiency will dictate the further uses of adhesives in demanding situations, but it is acknowledged that the specification and application of materials under site conditions is a major constraint. The increasing prefabrication of building elements and modules lends itself to the use of factory-applied adhesive with a greater degree of control. Making attachments between dissimilar materials, whether 'on-site' or 'in-factory', generally calls for the use of adhesive, often in conjunction with some form of mechanical interlocking. A very interesting case in point concerned the design of the IBM Travelling Technology Exhibition demountable pavilions by Ove Arup and Partners around 1984 (Figs. 1.6 and 1.7). The structures incorporated a mixture of aluminium, carbon-fibre, glulam, polycarbonate and stainless steel, with bonded connections being made between aluminium/glulam finger joints and between stainless steel plates attached to the polycarbonate skin.

13

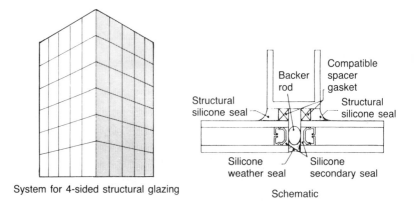

System for 4-sided structural glazing

Schematic

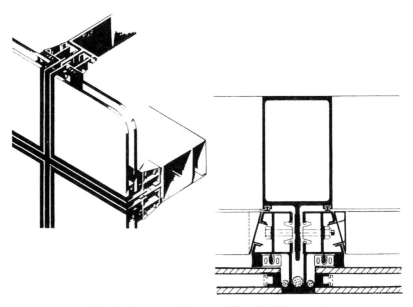

Detail of commercial system (Allan H. Williams Ltd)

Fig. 1.3. Four-sided structural silicone glazing.

Civil engineering

Many of the comments about the use of adhesives in building apply equally to civil engineering. The section on Historical Development outlined some of the uses of natural adhesives, for joining wood or

Fig. 1.4. Barnett Bank Centre, Miami, Florida; 15 floors, 4-sided structural silicone glazing.

material such as asphalt or flexible road and runway pavements. Since the 1950s synthetic polymer materials such as PVACs, SBRs and acrylic polymers have been used as general purpose bonding agents and adhesives, for example to improve the adhesion of surfaces. Whilst unsaturated polyesters, polyurethanes and acrylics all have their place among the applications for thermosetting adhesives, epoxy resins remain the major candidate materials for use in essentially non-structural situations such as:

(1) industrial flooring in the form of either pourable self-levelling systems or trowelled filled compositions.
(2) water-proofing membranes on concrete bridge decks
(3) skid-resistance layers on roads and other surfaces

Fig. 1.5. Brickell Bay Office Tower, Miami, Florida; 28 floors, 4-sided structural silicone glazing.

(4) resin mortars or concretes for expansion joint nosings on bridge decks, as a seating for bearing pads, and as a material for the surface repair of spalled concrete

(5) low viscosity formulations for the injection and sealing of cracks in concrete

(6) bonding new concrete to old, for the extension or repair of existing structures

(7) bonding of road marker studs to asphalt surfaces

(8) grouts for soil and rock consolidation

(9) epoxy coated bar reinforcement in structural concrete

(10) epoxy powder coatings as a protective layer to steel pipelines

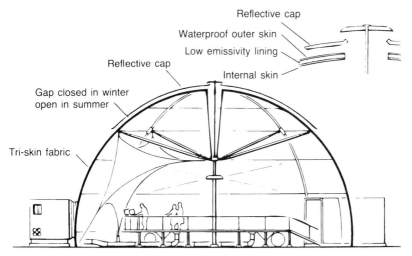

Fig. 1.6. IBM's travelling technology exhibition pavilion (courtesy Ove Arup & Partners).

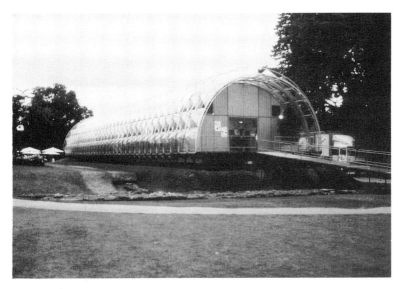

Fig. 1.7. IBM's travelling technology exhibition pavilion (courtesy Ove Arup & Partners).

17

Several applications utilise adhesives in a semi-structural manner in which the stresses to be resisted are dominantly compressive. These include:

(11) wire, rope or strand anchors
(12) steel fixings in concrete, rock or masonry
(13) self-levelling epoxy grouts for the support of heavy machinery, and as a bedding medium (resistant to indirect tensile forces) for crane and other rails
(14) segmental precast prestressed concrete structures such as bridges, in which epoxides have been used for nearly 30 years as a stress-distributing waterproof medium in joints.

Many of these applications will be considered further in Part 2 of the book, as will civil engineering uses of adhesives in a truly structural sense. By the latter it is implied that the adhesive is used to provide a shear connection between similar or dissimilar materials which enables the components being bonded to act as a composite structural unit. Within this definition, the main applications are:

(15) 'glulam'
(16) bonded external plate reinforcement for strengthening existing concrete structures
(17) bonded composite steel/concrete bridge decks
(18) structural steelwork connections
(19) sleeved steel bar and rebar connectors

A casein-adhesive-bonded arch made in 1910 supports the main roof of Oslo railway station to this day, and glulam is now a well established material for the construction of timber portals in buildings and for curved beams in roof structures and footbridges. It is often overlooked that glulam members of large cross-section have far greater fire resistance than corresponding ones of steel or concrete, and merely char as though composed of solid homogeneous timber. There is a steady demand, albeit limited in the UK, for concrete bridges and buildings to be strengthened using external steel plate reinforcement bonded selectively to their tension or shear surfaces (Fig. 1.8). Adhesives have been used to form the shear connection between steel and concrete in deck slabs in composite bridges, bonding to both (hardened) precast slabs and to fresh concrete. Such concepts have aroused interest in the US for the replacement of decks in highway bridges suffering from rebar corrosion. The use of adhesives to provide truly structural connections in steelwork is

18

Fig. 1.8. Strengthening of bridge structure with externally bonded steel plate reinforcement.

very much in its infancy, whilst rebar connectors are still under development.

Marine and offshore

Casein and then formaldehyde resin compositions have been used as adhesives and gap fillers in wooden boat construction for many years. The most significant use of resorcinol-formaldehyde resins was for the construction of wooden minesweepers for the Royal Navy, each vessel requiring some 3 to 5 tonnes – used mostly for laminating the hull. Polyester resins were then introduced to the marine industry and in 1950 the Scott-Bader Company recorded the construction of the first polyester-glass vessel. Intense development led to virtually all boat-builders producing moulds or finished GRP (glass reinforced plastic) craft, with wooden boat construction assuming a minority or specialist role. Many GRP-hulled boats, both naval and civilian, now rely significantly on resins for laminating, stiffening, the fabrication of sandwich panels, and for bonding attachments. Indeed the fitting out of many vessels is conducted with the large-scale use of non-structural sandwich- and insulating

19

panels using various bonded skin and core combinations. Epoxides were introduced into the industry for a range of bonding and gap-filling applications, one of the latter being for accommodating the tolerances and consolidating the bearings in large mooring buoys for oil tankers.

The most significant marine use of resins is actually in the form of paint corrosion protection systems for hulls. These include polyurethane and epoxide systems, the latter giving good alkali and solvent resistance in addition to providing superior adhesion to most substrates. Such systems take the form of zinc-rich epoxy and epoxy coal-tar combination hull paints. Epoxy powder coatings are also commonly used for the protection of steel pipelines, both on land and offshore.

A number of interesting structural applications of adhesives lie with the development of bonded stiffened plate structures for hulls, the development of lighter-weight composite superstructures (by bonding fibre-reinforced plastics to steel portals and frames), and with the repair of aluminium superstructures. The latter application arose because fatigue cracking developed in Type 21 Frigates, which was difficult to stop from propagating further by such means as drilling out the crack tips. Instead steel plates, up to 6 mm thick, were bonded over the cracks using a two-part cold-curing epoxide; carbon fibre laminate material was later used in place of the steel. The technique provides a rapid repair method with sufficient strength to contain cracking and minimise water leakage until such time as major replating can be carried out. The possibility of developing the procedure to provide sufficient durability and integrity as a permanent solution is being investigated by the Admiralty.

Many offshore steel structures are subjected to major damage due to accidents and collisions, or through stress fatigue failure of welded joints. As well as damage repair there are instances where it is necessary to modify or to strengthen existing structures. Conventional modification or repair techniques, often involving underwater welding, are extremely expensive and the development of the technique of underwater bonding of steel substrates represents a major technical advance in recent years. Adhesive-assisted repair methods for submerged steel structures have been developed by the Admiralty in conjunction with the Department of Energy, together with industry. This has required the formulation of hydrophobic filled cold-curing epoxies, as well as a sacrificial pretreatment technique; an adhesive-compatible hydrophobic film is deposited on the surface to be bonded which is then absorbed, or displaced, by the adhesive and enables adhesion to be gained underwater.

Another major offshore application, albeit still potential, lies with the stiff lightweight adoption of aluminium and/or polymer composite topside structures in order to reduce weight. Structural aluminium sections may be created by bonding together individual extrusions, and a most convincing demonstration of the potential has been developed by British Alcan (see Chapter 8).

Railway

Adhesives are used in a civil engineering context, in the track, and for the construction of rolling stock. Taking these applications in order, heavily filled acrylic adhesives have been used to provide a longitudinal shear connection between adjacent precast concrete beams in bridge structures. Although in the final deck mechanical devices aid the adhesive, bonding alone has been relied on when lifting pairs of beams into place. Resin injection has also been extensively used for the repair of brick and masonry arch bridges subjected to damage from vibrational or mortar loss.

In the track itself adhesives have been used in conjunction with bolts in the fishplate connections of continuously welded track. The adhesive serves primarily as an insulator, although it must also contribute to the 70 tonne axial loads and the bending arising from 25 tonne axle loads which the joint has to resist. In paved track, rail cleats have been glued into the supporting concrete raft.

In rolling stock adhesives have not been used for primary structural connections, but there are many examples of their usage in substructures. Amongst these are:

(1) aluminium skin and honeycomb structures, bonded with heat-cured film adhesive, for the sliding doors of suburban coaches
(2) attachment of top hat stiffeners to aluminium ceiling panels in inter-city coaches (Fig. 1.9)
(3) body side doors on the Class 58 locomotive are manufactured in zinc coated steel (IZ) and assembled using a two-part pre-mixed acrylic
(4) manufacture of suburban multiple unit coach bodyside panels consisting of 2 mm thick IZ steel skins and top hat section stiffeners. Advantages over conventional spot welding are those of time savings and an improved surface finish

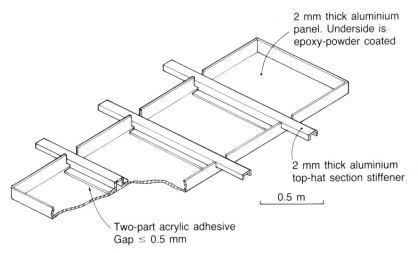

2 mm thick aluminium
panel. Underside is
epoxy-powder coated

2 mm thick aluminium
top-hat section stiffener

0.5 m

Two-part acrylic adhesive
Gap ≤ 0.5 mm

Fig. 1.9. Bonded aluminium ceiling panels for British Rail's Inter-City coaches.

(5) toughened epoxies have been used to construct a GRP box structure with internal foam reinforcement for Mk III inter-city coach doors.

In all of these applications the adhesive bond must remain sufficiently strong and stiff over the 20–40 year life of the rolling stock; the adhesive itself must also remain sufficiently flexible to allow joint deformation under impact loading. In these respects British Rail have favoured two-part acrylics, at least for use with zinc-coated steels.

Vehicle and automotive

Adhesives have been employed in the automotive industry since its earliest beginnings, with the use of natural resins to bond wood and fabric bodies. With the advent of the all-steel monocoque adhesives disappeared from the body, although they continued to be used for interior trim applications and are still used in this role today. The early metal-to-metal adhesives represented an extension of the low strength gap-filling interweld sealers and contributed little, if anything, to the overall strength of the assembly. However recent developments in synthetic resin technology have resulted in a very wide range of adhesive materials available to the design engineer, ranging from weak low modulus ones to tough high strength formulations. Adhesives are now designed into vehicles at the concept stage, and will be used increasingly in demanding roles as design philosophies, and the materials from which vehicles are constructed, evolve.

Early metal applications in car body manufacture were concerned with the bonding of stiffeners to bonnet and boot lids in order to overcome problems of flutter and rattle at high speed. These components, together with flange seams and clinched seams in door and roof assemblies, remain popular candidates for adhesive bonding (Fig. 1.10). The adhesives are generally applied to steel contaminated with mill oils at the body-in-white stage, and a number of tack welds may additionally be used to hold the assembly in place. The use of robotic automatic dispensing equipment has enabled a consistent application quality to be achieved for mass production, although careful controls are necessary to monitor the process and the materials used.

Adhesives are used in many other applications on vehicles, from attaching trim, for thread-locking, for gasketing in the engine, and for components such as headlamps, radiators and brake linings. The use of polyurethane adhesive-bonded direct glazing for the front, rear and side screens of cars is now a well established technique, contributing to both a flush aerodynamic exterior as well as enhanced body stiffness. One recent innovation has been developed by Peugeot in which a carbon fibre drive shaft is bonded into steel yokes at either end to pick up the transmission and drive. The carefully shaped co-axial joint design represents a development which has been used successfully in rally cars and latterly in standard production models by other French motor manufacturers.

The future use of adhesives in motor vehicle body shell construction depends significantly upon the materials used. Aluminium may compete with high strength low alloy steels, and reinforced plastics will increasingly be used in combination with other materials. The BL ECV3 concept vehicle was one demonstration of a lightweight vehicle comprising all three materials, in which very significant use of adhesive bonding was implemented. Currently many specialist vehicles are constructed with mixtures of materials, particularly plastics, in which large volumes of adhesives are used. At the same time mass-produced vehicles are increasingly being fabricated with coated steels for which joining by adhesive bonding is very attractive in order to minimise damage to the surface coating.

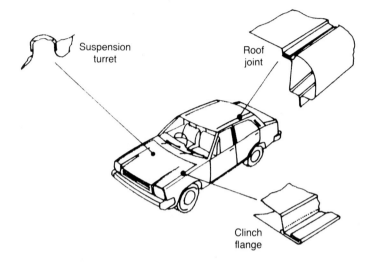

Applications of structural epoxy adhesives

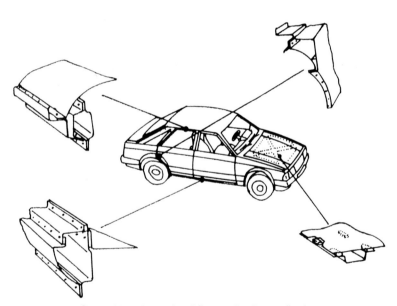

Spotweld sealer and anti-flutter adhesive applications

Fig. 1.10. Typical adhesive applications in car bodies.

Adhesives are used extensively in the commercial vehicle sector, particularly for the joining of dissimilar materials. For instance, many truck cabs now comprise large amounts of reinforced plastic bonded steel. Container vehicles, refrigerated or otherwise, are often constructed from bonded sandwich panel assemblies; further, by making the bonded container act structurally much of the conventional chassis is unnecessary, so reducing weight or enabling a higher payload to be transported. An additional approach to reducing the structural chassis weight has been developed by Leyland–Daf in conjunction with British Alcan. The Leyland TX450 concept vehicle (Fig. 1.11) comprises an aluminium monocoque tubular chassis of great torsional stiffness. The hollow longitudinal members are formed from bonding three individual aluminium extrusions together, whilst the cross members also employ adhesive bonding. A weight saving of over 30 per cent has been realised when compared to a traditional steel ladder frame chassis. Naturally, careful provision for surface treatment of the aluminium for bonding has to be made, as is the case in the development of aluminium-bodied cars.

Vehicle repair techniques, overseen by the insurance industry, are also being developed for bonding sheet steel panels. Epoxides and polyurethanes are used, and the benefits of using adhesives include the need for access only to the outside surface of the repair area as well as overall cost and time savings. Crash and impact tests have confirmed the suitability of such repairs as well as highlighting, as with new build operations, the need for judicious use of tack welds or rivets.

1.5   Relevance to civil engineering

We have seen that structural adhesive bonding has been employed in the aerospace and other industries for several years, whereas its use in civil engineering is relatively recent. Adhesive bonding can present some very unfamiliar concepts to many engineers and, in particular, civil engineers and this may partially explain the extremely cautious embrace of the technology. Thus it is perhaps worth considering at this stage the applicability of the experience gained in industries such as aerospace and automotive to future applications in civil engineering.

Projects in the construction industry are distinguished from those

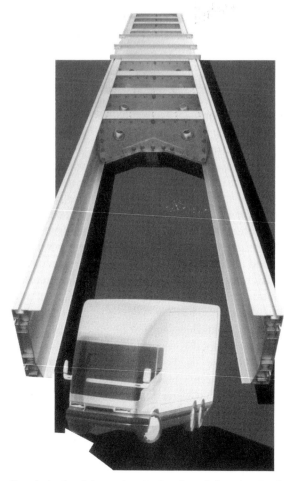

Fig. 1.11. Bonded aluminium chassis developed for the Leyland TX450 truck (courtesy British Alcan Aluminium).

in other industries by the fact that they are generally unique, are specifically commissioned and are built *in situ*. Unlike motor cars and aeroplanes, civil engineering structures are not generally mass produced, they are not built first and sold later and, with the exception of pre-fabricated components, cannot be assembled economically under cover. These factors all have an influence on the relevance of previous adhesive research and practical experience.

At the design and specification stage there exist few design tools and recommendations available for general applications of adhesives.

The civil engineer often requires an adhesive to perform a gap-filling role as well as being a stress transfer medium, with the adhesive often being used in thick bondlines. Far lower levels of dimensional precision would also be anticipated in construction, which further compounds the problem. This immediately calls into question the validity of any existing design tools developed for engineering applications generally involving the use of adhesives in very thin or thin bondlines.

The choice and availability of adhesives for Civil Engineering applications is also somewhat limited, because manufacturers have naturally tailored their products to the needs of the aerospace and general manufacturing industries. Thus products developed specifically for bonding steel and concrete are currently few and far between.

The bonding operation itself can be fraught with difficulties. The outdoor nature and scale of construction projects means that surface treatment techniques need to be appropriate to the practical conditions. Simple grit blasting and degreasing methods may even be very difficult in remote locations or in adverse weather conditions. Any form of quality structural bonding requires close control and protection from the weather. It may be that pre-treated or pre-primed components (like coil-treated sheet aluminium) could be used which present a reproducible surface to adhesives. Alternatively the development of sacrificial pretreatment technology, used for underwater bonding, may have considerable merit for certain situations if heat-cured adhesive products cannot be used. The application of adhesive materials is unlikely to be possible by automated equipment, although automatic mixing and dispensing may be practical. In many instances products which cure at ambient temperature will have to be used, although at temperatures below 10 °C the rate of curing drops off very rapidly; this may imply the need for space heating. If the superior performance of heat-cured adhesives is required then it may be possible to provide the necessary curing regime over small areas by using electrical heating tapes and blankets, infra-red heaters or even induction heating methods.

Many civil structures are designed for lives in excess of one hundred years and during this time they will be subjected to extremes of temperature, moisture and probably sustained as well as cyclic loading. The operating conditions in other industries may be quite different, in particular the life span of a vehicle or aircraft is unlikely to exceed twenty years. The engineer will need to satisfy himself that the adhesive joint remains capable of performing its intended role for a specified time; for primary structural connections this will usually be the design life of the structure. In repair or strengthening applications a shorter time may be acceptable in the knowledge that the process can be repeated if necessary at periodic intervals. The ability to check the integrity of bonded joints is very important although non-destructive methods are not yet generally available or applicable for many current and potential bonded joint applications in civil engineering.

1.6 Closing remarks

Much pioneering research work has been completed in the science and technology of adhesion and adhesives. A lot of that work has been directed to solving problems for technically demanding bonding applications, and the resultant understanding has enabled further development and technology transfer. Whilst the substrates to be bonded, the adhesives used, the operating conditions experienced and the rigour of control which can be exercised within the construction industry are often very different from those occurring in other industries, some useful insight and knowledge can be accumulated from examining the experience in other industries. Undoubtedly there is much to be gained which is applicable to off-site prefabrication of components, particularly for building construction applications. However the activities and applications in the various other industries that have been reviewed briefly in this Chapter serve to indicate current usage and R&D for technology transfer consideration into future civil engineering applications. The more widespread application of adhesives to bonding civil engineering structures remains both a matter of time and a matter of example. It must be concluded that further exploitation awaits a greater total documentation of design approaches for typical structural details, as well as the incorporation of design guidance in Codes of Practice.

1.7 Further reading

Adams, R.D. and Wake W.C. *Structural Adhesive Joints in Engineering*, Elsevier Applied Science, London, 1984.
Aitken, D.F. (Ed), *Engineer's Handbook of Adhesives*, The Machinery Pub. Co., Brighton, 1972.

Further reading

Adhesives, Sealants and Encapsulants (ASE) Conferences, London, November 1985 and 1986. See papers on applications in the proceedings from Network Events, Buckingham.

Adhesives, Surface Coatings and Encapsulants (ASE) Conference, Brighton 1988. Proceedings from Network Events, Buckingham.

Engineering Applications of Adhesives, Seminar, London, 1988. Papers from Butterworth Scientific, Guildford.

Feldman, D. *Polymeric Building Materials*. Elsevier Applied Science, London, 1989.

International Journal of Adhesion and Adhesives, Special Issues: 1 – Adhesives in Civil Engineering, **2**, No 2, 1982; 2 – Adhesives in Land Transport, **4**, No 1, 1984.

Jackson, B.S. (Ed), *Industrial Adhesives and Sealants*, Hutchinson Benham, London, 1976.

Kinloch, A.J. (Ed), *Developments in Adhesives – 2*, Applied Science Publishers, London, 1981.

Lees, W.A. *Adhesives in Engineering Design*, The Design Council, Springer-Verlag Publishers, London, 1984.

Lees, W.A. *Adhesives and the Engineer*. Mechanical Engineering Publications Ltd. London, 1989.

Mays, G.C. *Materials Science and Technology*, **1**, 1985, pp. 937–43.

Plastics and Rubber Institute, Adhesives Group, *Symposium on Adhesives and Sealants in Building and Construction*, London, 22 February 1988.

Structural Adhesives in Engineering (SAE) Conference, Bristol University: SAE I, July 1986: Proceedings from Mechanical Engineering Publications, Bury St Edmunds; SAE II, September 1989: Proceedings from Butterworth Scientific Conferences, Guildford.

Skeist, I. (Ed), *Handbook of Adhesives*, Van Nostrand Reinhold Co. New York, 1990.

Wake, W.C. (Ed), *Developments in Adhesives – 1*, Applied Science Publishers, London, 1977.

CHAPTER TWO

Adhesive classification and properties

2.1 Engineering and non-engineering adhesives

Adhesives may be classified as either organic or inorganic materials in a number of different ways; for example by origin, by method of bonding, by end use or on a chemical basis (1). Table 2.1 gives a broad classification of the organic adhesives based upon origin under the general headings of animal, vegetable, mineral, elastomeric, thermoplastic and thermosetting adhesives.

Animal glues are generally based on protein either in the form of mammalian collagen, from fish or from milk. They tend to be 'sticky' which is useful for applications requiring an instant grab or bond.

The common vegetable glues are based on either starch or cellulose. Unmodified starch dispersed in water is used to form paper pastes but a large proportion is now used in modified forms such as dextrin. Cellulose-based glues are produced by reacting hydroxyl groups present in the polymer chain with different reagents to form a variety of adhesives.

Mineral adhesives include silicates and phosphates for high temperature use and naturally derived products such as bitumen and asphalt.

The elastomeric group of adhesives is based on natural rubber latex and its derivatives or totally synthetic rubber known as SBR (styrene butadiene rubber). There is now a wide range of synthetic rubber adhesives based upon SBR including nitrile and butyl rubber. Another elastomeric adhesive is the versatile polyurethane rubber group.

Thermoplastic adhesives are so called because they may be softened by heating and rehardened on cooling without undergoing chemical changes. There is a wide range of such adhesives many of which contain the vinyl group. The PVA (poly vinyl acetate) group of general purpose adhesives which are formulated in aqueous emulsions tend to be moisture sensitive. Others are derived from

Table 2.1. *Adhesive classification*

Group	Type	Source	Use
Animal	gelatin	mammals, fish	can labels
	casein	milk	plywood, blockboard
	albumen	blood	
Vegetable	starch	corn, potatoes, rice	paper, packaging
	cellulose acetate ⎫	cellulose	leather, wood, china
	cellulose nitrate ⎬		
Mineral	asphalt/bitumen	earth's crust	road pavements
Elastomeric	natural rubber	tree latex	carpet making
	SBR	synthetic	tyre vulcanising
	nitrile rubber	synthetic	PVC solvent glue
	polyurethane rubber	synthetic	fabrics, bookbinding
	silicone rubber	synthetic	
Thermoplastic	PVA	synthetic	wood and general
	polystyrene	synthetic	model making
	cyanoacrylates	synthetic	plastics, metals, glass, rubber
	liquid acrylic	synthetic	structural vehicle assembly
Thermosetting	phenol-formaldehyde ⎫	synthetic	chipboard and plywood
	urea-formaldehyde ⎬		
	unsaturated polyesters	synthetic	glass fibre, resin mortars
	epoxy resins	synthetic	structural, especially metal to metal
	polyurethane	synthetic	semi-structural uses with plastics, metals, wood and sandwich panel construction

polyesters, nylons and, importantly, the cyanoacrylates. The latter group set very quickly when squeezed out in thin films. These monomers are liquids of low viscosity which polymerise very easily on contact with traces of moisture invariably present on the surface of the adherend. Their main disadvantages with regard to potential structural use, apart from the very rapid set, are a lack of moisture resistance when used to bond metal surfaces and their restriction to thin bondlines.

The cyanoacrylates are only part of the wider range of acrylic adhesives now on the market. The development of liquid methacrylates and acrylates followed from the introduction of anaerobic gasketing sealants and thread couplants. Two-part liquid acrylic adhesives are now available in forms which allow the liquid to be applied to one surface and the initiator (activator or catalyst) to the other. Contact during assembly produces a strong bond although the glue-line thicknesses necessary to achieve intimate mixing on contact may be rather too thin for civil engineering use. Pre-mixing the two components can result in a rapid set unless specially formulated. Some reservations have also been expressed regarding durability against moisture when used on metal surfaces, although there is conflicting evidence on this point. Nevertheless, the acrylics show potential for providing an alternative source of structural adhesive to the epoxy resin in the future, particularly as they are believed to be less toxic.

The molecular chains present in thermosetting adhesives undergo irreversible cross-linking on curing. Unlike thermoplastics they do not melt or flow on heating but become 'rubbery' and lose strength. Synthetic resins formed from a reaction between urea, phenol, resorcinol or melamine and formaldehyde are common adhesives of this type used in the production of glulam. However, for hot and moist exposure conditions phenol or resorcinol formaldehyde products only are favoured. The other very important thermosetting adhesives come from the epoxy and unsaturated polyester groups.

As structural adhesives, epoxies are the most widely accepted and used. They typically contain several components, the most important being the resin. To the base resin is added a variety of materials, for example hardeners, flexibilisers, tougheners and fillers. These all contribute to the properties of the resulting adhesive. Formulations may be further varied to allow for curing at either ambient or elevated temperatures. The epoxies and polyesters, together with acrylics, polyurethanes and synthetic polymer lattices will be discussed in greater detail in the sections which follow. However, it will become evident that epoxy resins have several advantages over other polymers as adhesive agents for civil engineering use, namely:

(1) high surface activity and good wetting properties for a variety of substrates
(2) may be formulated to have a long open time (the time between application and closing of the joint)
(3) high cured cohesive strength; joint failure may be dictated by adherend strength (particularly with concrete substrates)
(4) may be toughened by the inclusion of a dispersed rubbery phase
(5) lack of by-products from curing reaction minimises shrinkage and allows the bonding of large areas with only contact pressure (in stark contrast to the phenolics commonly used in the aerospace industry)
(6) low shrinkage compared with polyesters, acrylics and vinyl types; hence, residual bondline strain in cured joints is reduced
(7) low creep and superior strength retention under sustained load
(8) can be made thixotropic for application to vertical surfaces
(9) able to accommodate irregular or thick bondlines (e.g. concrete adherends)
(10) may be modified by (a) selection of base resin and hardener (b) addition of other polymers (c) addition of surfactants, fillers and other modifiers.

In practice, most commercially available epoxide resins are blended with a variety of materials to achieve desirable properties, as outlined above; indeed, the possibilities for useful new combinations are numerous. A major disadvantage of epoxides is that these various modifications and the materials concerned make them relatively expensive when compared with other adhesives.

2.2 Generic classification of adhesives

Epoxy resins

Epoxy resins have only been available commercially since the Second World War and are traditionally based upon the reaction of epichlorohydrin on bis-phenol A, to give a liquid compound of

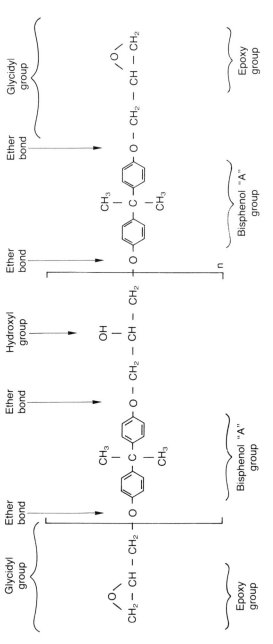

Fig. 2.1. Chemical structure of DGEBA (Ref. 2).

linear molecules terminating with epoxy groups and having secondary hydroxyl groups occurring at regular intervals. The formal representation of the chemical structure of the resulting diglycidyl ether of bisphenol A (DGEBA) is shown in Fig. 2.1(2). Thus, epoxy resins can be regarded as compounds which normally contain more than one epoxy group per molecule. A general categorisation of the base resins which may be encountered in epoxy adhesive formulations is given in Table 2.2(2) and, of these, the DGEBA and DGEBF resins, or blends of the two, are the types most commonly encountered in two-part room temperature cure epoxy adhesives used in the construction industry.

Adhesive properties of epoxies are obtained by polymerisation using a cross-linking agent, commonly referred to as the hardener,

Table 2.2. *A general categorization of epoxy resins (Ref. 2)*

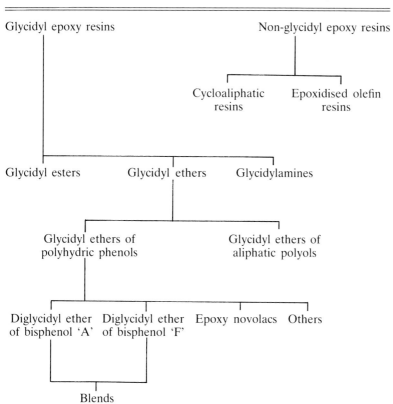

Table 2.3. *Hardeners for epoxy adhesives (Ref. 2)*

Polyamines   Polyamides   Polysulphides   Polythiols Polymercaptans   Dicyandiamides   Acid Anhydrides

Blends

- Aliphatic
  - Primary
  - Secondary
  - Tertiary
  - Modified
- Cycloaliphatic
- Aromatic

to form tough three-dimensional polymer networks. A general categorisation of the common hardeners is given in Table 2.3(2) and their general effect on adhesive properties will now be discussed.

Epoxy hardeners

Aliphatic polyamines. These are one of the most commonly used hardeners in room temperature curing epoxy adhesives. With glycidyl ether resins only relatively small quantities (6–12 parts per hundred of resin) are required, although sometimes the rate of reaction may be a little too fast for convenience, particularly in conditions of high ambient temperature. The hardened adhesives are rigid and provide good resistance to chemicals, solvents and water. Some aliphatic amine compositions react with carbon dioxide and moisture in the atmosphere causing a characteristic tacky surface layer on the cured product. For this reason they do not provide a good bonding surface for the addition of further epoxy resin compounds(3). All aliphatic amines are caustic and can be difficult to handle so that they are generally modified for commercial use.

The basic chemical reaction between an epoxy and a primary amine can be represented by:

$$-\overset{|}{C}-\overset{|}{\underset{\diagdown \diagup}{C}}- \quad + \quad \overset{|}{R}NH_2 \longrightarrow$$
$$O$$

$$HO-\overset{|}{\underset{|}{C}}-\overset{|}{\underset{|}{C}}-\overset{}{\underset{\underset{R}{|}}{N}}-\overset{|}{\underset{|}{C}}-\overset{|}{\underset{|}{C}}-OH$$

A common modification which is performed commercially to selectively improve the performance is to react a glycidyl ether resin with an excess of amine groups to produce a resinous amine adduct hardener. These have advantages over unmodified versions in that they result in more convenient mixing ratios, are less dangerous to handle and can exhibit reduced moisture sensitivity.

Cycloaliphatic amines. These can give good cures under adverse conditions of low temperature and high humidity and as such form the base of many modern grouts and resins used in civil engineering. Chemical resistance and mechanical properties are similar to those of the aliphatic amines but they are still skin irritants.

Aromatic amines. Most only give a sluggish cure at room temperature and were primarily developed to provide improved heat and chemical resistance. Room temperature cures can be obtained by dissolving the amine in a diluent but this tends to reduce the strength of the adhesive. The benzene ring structure of aromatic amines results in a dense stable cross-linked network which tends to give good water resistance but rather brittle cured resins and lower strengths than can be achieved with aliphatic amines. Their intrinsic brittleness may, however, be reduced by the addition of modifiers.

Secondary and tertiary amines. Tertiary amines can be used as sole curing agents but in adhesive formulations they are more normally found in blends, for example as accelerators with polyamide systems(4) and as an essential part of polysulphide systems. The cured resins tend to be less temperature and chemical resistant than the amines described above. Secondary amines react first like primary then as tertiary amines.

Polyamides. These should be strictly called polyamidoamines since it is still the amine group which provides the reactivity. However, larger quantities of hardener are required than with aliphatic amines and they tend to have a slower rate of reaction. Flexibility is generally improved but other properties tend to be inferior, although curing at elevated temperature will effect an improvement. Mixing ratios are less critical and polyamides are less likely to give skin irritation so that they are especially suitable for use in the twin-tube packages sold as all-purpose household glues. Water resistance may be unsatisfactory for longer term structural applications.

Polysulphides. Liquid polysulphide rubbers have been used as hardeners for epoxy resins for many years. However, they are generally used as blends with tertiary amines and will be considered further as flexibilisers.

Polymercaptans and polythiols. These form the hardeners for the quick setting epoxies produced mainly for the DIY market(5). As such they are little used in civil engineering.

Dicyandiamides. These are latent catalysts which are inactive at ambient temperatures but are released for reaction at about 90 °C; at temperatures approaching 150 °C and above they react rapidly. They are primarily used in single part epoxy adhesives and require storage in cool conditions in order to remain stable for a reasonable shelf life.

Acid anhydrides. Another group of hardeners which require high temperature cure schedules(6) but give improved thermal stability over amine cured products. Because of several drawbacks including brittleness they are not often used for adhesives.

The wide range of properties available from different combinations of epoxy resins and their hardeners can be further extended by the use of other additives. These include anti-oxidants, diluents, flexibilisers, stabilisers, tougheners, fillers, surfactants and adhesion promoters.

Epoxy additives

Fillers. In practice most epoxy resin systems have fillers incorporated, often simply to reduce cost although they may also assist in gap filling, reduction of creep, reduction of exotherm, corrosion inhibition and fire retardation. Their incorporation will also alter the physical and mechanical properties of the adhesive. Construction resins in particular often include a large volume fraction of sand or silica.

In general, fillers are inert materials which may be organic or inorganic in nature. They increase the viscosity of the freshly mixed system but some also provide a shear rate dependency referred to as thixotropy. This is particularly useful for adhesives which are to be applied to vertical surfaces so that run-off can be minimised. The addition of fillers also serves to reduce the exotherm and subsequent thermal shrinkage on cure, and to extend the pot life. Thermal expansion coefficients are also lowered. With particulate fillers tensile and flexural strengths are usually reduced but compressive strength is improved. However, fibrous fillers can improve tensile strength and impact strengths. The effect on moisture and chemical resistance is less clear, there being some evidence(27) that the presence of fillers can cause a wicking action aiding ingress of moisture. However, plate-like filler particles may serve to reduce water transport. Traditionally asbestos was commonly used as a filler but for health and safety reasons this has now been replaced by materials such as silica flour, talc and aluminium powder.

Diluents. These are generally incorporated to reduce the viscosity of the freshly mixed adhesive to offset the effect of the filler. This may be required to improve handling and spreading characteristics or to allow filler additions which tend to reduce cost. Other properties of the fresh and hardened adhesive can be affected by the use of diluents, for example pot life, flexibility and glass transition temperature. If the diluent is non-reactive, such as solvents which remain in the cured system, the net result is a deterioration of chemical and mechanical properties such as increased shrinkage and reduced adhesion. Reactive diluents containing epoxy compounds are capable of combining chemically with the resin/hardener system.

Flexibilisers. These are long chain molecules which may either cause a mechanical plasticising action (often referred to as plasticisers) or react to some extent with the resin during cure to increase flexibility by basically neutralising the attraction between adjacent chains. They are used to improve the impact resistance, peel strength or flexibility of epoxies but can cause side effects such as reduced tensile strength and transition temperature. They may be resinous in nature or be derived from curing agents such as polysulphides.

Tougheners. Unmodified epoxy systems tend to be strong in shear, compression and tension but brittle when cleavage or peel forces are imposed (Fig. 2.2). In general, adhesive joints are designed to avoid the latter forces but in practice they can rarely be eliminated entirely. Flexibilising can produce improvement in these properties at the expense of cohesive strength but in recent years a technique known as 'toughening' has been developed to overcome this problem. Toughening is achieved by the inclusion of a dispersed rubbery distortable phase within the load-bearing glassy matrix of the adhesive (Fig. 2.3). The aim is to provide a physically separate but chemically linked zone which absorbs fracture energy and prevents crack propagation. The analogy with drilling out the crack tips in metal is often drawn. The main energy absorption mechanisms are shear yielding in the matrix, crazing in the matrix and deformation

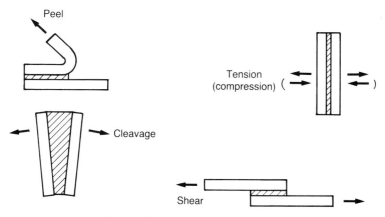

Fig. 2.2. Types of in-plane stress in a bonded joint.

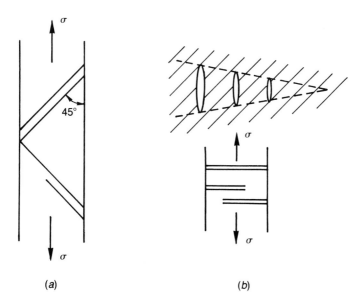

Fig. 2.3. Schematic representation of failure mechanisms in rubber tough-ened adhesives. (*a*) Shear yielding. (*b*) Crazing (not necessarily applicable to epoxies).

and failure in the rubber particles (Fig. 2.3). A common toughening agent is liquid carboxyl-terminated butadiene acrylonitrile (CTBN) rubber which maintains the stiffness, hardness and temperature resistance of the adhesive much better than liquid plasticisers or flexibilisers. Both two-part and single-part epoxies can be toughened

41

in this way(7), although it is more difficult to achieve a well dispersed rubbery phase in polymers cured at ambient temperature.

Adhesion promoters. Sometimes referred to as coupling agents, these additives have the ability to enhance resin adhesion to surfaces such as glass or metals(8). The most popular type are silanes which can either be mixed with the adhesive itself or applied to the substrate as a primer. They will be further considered in Chapter 3 under adhesion and surface pretreatment.

Commercially available resins may also contain fire retardants, anti-oxidants, surfactants, and so on. It may thus be deduced that useful adhesives are complex and sophisticated blends of many components. Not only the choice of hardener but also the presence of additives may affect significantly the physical and mechanical properties of both the freshly mixed and hardened adhesive. A model which is often used to represent the structure of thermosetting cross-linked adhesives is shown in Fig. 2.4(9). With epoxies the resin is represented by 'hooks' of high molecular weight and the hardener by the 'eyes' of short chain cross-linking segments(10). An increase in hook or eye length or a decrease in the number of linkage points will result in an increase in elongation, peel strength, impact strength, low-temperature performance and permeability to water; it will also result in a decrease in elastic modulus, hardness, thermal stability, hot strength and resistance to chemical attack.

The rate of cure is temperature dependent and many formulations stop curing altogether below a temperature of about 5 °C. If carefully formulated the change in volume between the uncured resin-hardener system and the fully cured polymer can be very low. This property, together with their relatively high strength and claimed resistance to moisture and chemical attack, forms the basis of the use of epoxy resins as structural adhesives.

Polyesters

Whereas in epoxies the 'hooks' are provided by the resin and the 'eyes' by the hardener, in polyesters both 'hooks' and 'eyes' are located in the unsaturated resin(9). The addition of a curing agent or catalyst based on organic peroxide initiates the chemical reaction and promotes cross-linking within the resin. Unsaturated acids are

Generic classification of adhesives

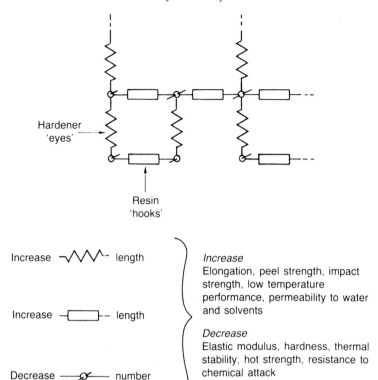

Fig. 2.4. Diagrammatic model of a thermoset adhesive (Ref. 9).

thus cross-linked by interpolymerisation with styrene to produce the so-called 'unsaturated polyesters' used as the basis of glass reinforced plastics and adhesives for constructional purposes.

Unfortunately, with polyester resins the contraction during cure can be as high as 10% by volume. This is partly due to the high level of thermal contraction which, unlike epoxies, tends to occur after the resin has set and can result in stress build up, possibly accompanied by interfacial cracking, at the resin/substrate boundary. In addition, there is a volume change during the transition from the uncured liquid phase to the hardened resin resulting in further curing shrinkage, in some cases due to solvent loss. Thus, there are usually strict limitations on the volume of materials that can be mixed and applied at any one time.

Reservations have also been expressed regarding the suitability of polyesters as structural adhesives because of the poor creep

43

resistance under sustained load and their bonding efficiency in damp or wet conditions, particularly to alkaline substrates. Nevertheless, they may be useful materials if a very rapid gain in strength is required from a material with a reasonable usable life after mixing. Several formulations also have the advantage of being able to cure in sub-zero temperatures.

Acrylics

The liquid acrylics form a further group of unsaturated reactive resins and these are now available as two-part mixed or unmixed products. Compared with polyesters they are a relatively recent addition to the range of adhesives potentially suitable for structural joints. Many are based on the monomer methylmethacrylate which is polymerised by the addition of a small quantity of initiator or hardener.

In the two-part mixed versions the amount of hardener added to the monomer can vary between 5% and 50% depending on the formulation. These are suitable as general purpose adhesives capable of filling gaps up to 5 mm in thickness with a usable life of between 10 and 60 minutes. In the unmixed versions the hardener is applied as a thin film to one of the surfaces to be bonded. The monomer is then applied to the opposite surface, the parts immediately fitted together and held under contact pressure. The polymerisation starts instantaneously after the adhesive touches the hardener and requires about 15–20 minutes before the bonded joint can be handled. Although the overall performance is generally better than with the mixed versions, the gap which can be filled is limited to about 0.5 mm.

Many formulations are claimed to be capable of absorbing normal residues of rolling oil on steel surfaces, thus reducing the surface preparation requirements as compared with, say, epoxies. As with epoxies, acrylics can be toughened by the introduction of a rubbery phase within the matrix of the adhesive and they are particularly suitable for polymeric adherends because the surface is compliant.

Three further points are worthy of mention. Firstly, the methylmethacrylate monomers are highly inflammable with a flash point at about 10 °C. Secondly, oxygen can act as an inhibitor of polymerisation so that the exposed edges of a joint, particularly with thicker bond lines, may never fully cure and may lead to concern over durability along the joint perimeter. Finally, the best acrylics have a distinctive odour and attempts to overcome this invariably lead to a reduction in overall performance.

Polyurethanes

The original development of polyurethanes was relatively slow due to early competition from more favourable epoxy formulations. However, in the 1950s and 1960s work on surface coatings suggested that polyurethanes were capable of being formulated to achieve a wide variety of performance properties, some of which were appropriate to applications as adhesives. Nowadays uses are wide-ranging, taking advantage of strength in automotive panel bonding, through to flexibility and toughness required in shoe manufacture. Building and construction applications to date have been semi- or non-structural, for example in prefabricated building elements, roofing and flooring, and as movement joint sealants.

Urethanes are manufactured from iso-cyanates and it is the iso-cyanate groups which are capable of reacting with any material containing reactive hydrogen, to form urethane linkages. Thus linear chain molecules can be built up by reaction with water, amines or alcohols. A wide variety of formulations can result including elastomers, two part cross-linked urethanes, single part moisture curing products, hot melt and water borne urethanes. The 100% solids systems hold the most promise for structural bonding since they do not involve the loss of solvents or water during the cure process. Current applications include sandwich panel construction, and the bonding of polymer composites, metals and timber. The bonding of highly alkaline substrates such as concrete is not however advised.

Polymer bonding aids

Although not strictly used as prime structural adhesives, polymer latices or dispersions have been included here because of their increasing use as aids to bonding in the patch repair of spalled concrete. They usually take the form of a polymer–cement slurry which is applied to the moistened and prepared concrete surface. In general, the repair mortar must then be applied before a film forms on the surface of the bonding aid. Many cementitious repair mortars themselves are today modified by polymers of similar generic type to those used in bonding aids.

Before discussing the various forms of polymers which may be encountered it is pertinent to describe briefly the basic structure of water-based polymer dispersions(11). The starting point is a monomer which forms droplets in water. Aqueous surfactants are adsorbed at the droplet surface to stabilise the emulsion before an initiator is added to cause polymerisation under controlled conditions of pressure, temperature and stirring rate. The latex is thus an aqueous dispersion of small discrete polymer particles to which is added a range of additives, such as coalescents, anti-foaming agents, bacteriocide and anti-oxidants, to improve shelf life and properties related to its end use.

Most bonding aids are supplied for civil engineering use at a total solids content of about 50% by weight. The materials must be formulated to meet the particular requirements for concrete repair, that is to be stable in a relatively high alkaline environment and to have a minimum film-forming temperature of 5 °C. Film-forming properties are particularly important but the process is complex in polymer dispersions. Essentially it occurs as a result of water evaporation which eventually leads to particle contact. However, the stabilising surfactant system employed and other properties such as particle size, surface tension and viscosity will also influence film formation. As with all the previous adhesive systems which have been discussed the various additives incorporated into such bonding aids can have as much influence on the final properties as can the nature of the polymer itself.

Thermoplastic polymers supplied for use as concrete bonding aids include polyvinyl acetate (PVA), acrylics, styrene acrylates and styrene butadiene rubber (SBR). The PVAs tend to be used for non-structural applications, for example bonding render to concrete, as they tend to be moisture susceptible. Acrylics and styrene acrylates exhibit superior resistance to water and can be nearly as good as SBRs in this respect. There is also some evidence to suggest that the acrylics are less sensitive to premature film-forming in drying conditions than are the SBRs. Indeed, the Department of Transport in their specification on concrete bridge repair(12,13), restrict the use of polymer bonding aids to acrylic systems. Counter to this is the much longer and generally successful track record of SBR cement bonding agents.

The neat or diluted latex can be used without cement but can only be applied in relatively thin films, resulting in a lack of tack and lower tensile strengths than systems containing ordinary portland cement. Latices combined with cement, typically in ratios of between 1 : 1 and 1 : 2 by volume, result in reasonably resilient yet strong bond coats. The latex/cement mixture achieves its desirable properties of substrate wetting, substrate adhesion, strength and resilience from the way in which the polymer latex particles coalesce on drying to form continuous strands which penetrate the cement matrix and bridge voids between cement particles and the substrate. In addition the latex improves workability and enables a reduction in the water : cement ratio to be made.

2.3   Adhesive properties

The engineer will be concerned with the behaviour and performance of the selected adhesive from the time he first purchases it from the manufacturer, through the mixing, application and curing phases to its properties in the hardened state within a joint over the intended design life. Thus the properties of interest in approximate chronological order are likely to include:

(1) Unmixed   – shelf life

(2) Freshly mixed   – viscosity
                     usable life
                     wetting ability
                     joint open time

(3) During cure   – rate of strength development

(4) Hardened   – strength and stress/strain characteristics
                     fracture toughness
                     temperature resistance
                     moisture resistance
                     creep
                     fatigue

To a large extent the performance of the hardened adhesive within a joint will be discussed in considerable detail in Chapters 3 and 4. Thus, the purpose of the review which follows is to compare the basic properties of bulk adhesive from the time of mixing to the hardened state.

(1) Unmixed state

Shelf life – this is the period for which the unmixed adhesive components may be stored without undergoing significant deterioration. Most structural adhesives have shelf lives in excess of six months and maybe up to several years. Shelf life is often extended by storage at low temperature, for example in a refrigerator. It is usually advisable to stir each adhesive component before use as sometimes segregation can occur during storage. In so doing care should be taken to avoid the entrapment of air.

(2) Freshly mixed state

Viscosity – the shear stress, τ, required to displace an element of fluid as shown in Fig. 2.5 is directly proportional to the shear rate $D = \mathrm{d}v/\mathrm{d}y$ where v = velocity. The constant of proportionality is known as the coefficient of viscosity, η, and fully defines a Newtonian fluid (see Fig. 2.6). However most structural adhesives exhibit pseudo-plastic characteristics together with a yield stress as shown in Fig. 2.7. This implies that they have gap filling properties as well as good workability. Typical flow curves determined at a shear rate of 10^{-1} s and 20 °C for a range of epoxy adhesives are illustrated in Fig. 2.8(14). Apparent viscosities range from 30 to 260 Pa s and yield stresses from 0 to 70 Pa. For ease of spreading, a viscosity within the range 20 to 150 Pa s at a shear rate of 10^{-1} s is desirable. For spreading on vertical surfaces a yield stress of at least 20 Pa is essential. Changes in ambient temperature and the inclusion of fillers can significantly alter flow properties as illustrated in Fig. 2.9(14).

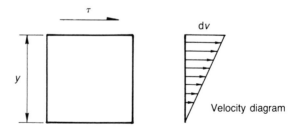

Fig. 2.5. An element in a Newtonian fluid.

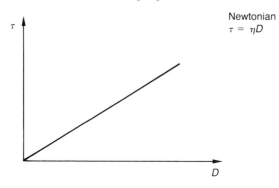

Fig. 2.6. Newtonian fluid.

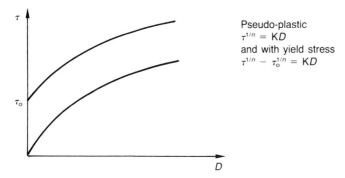

Fig. 2.7. Pseudo-plastic fluid with yield stress.

Adhesives are sometimes described as thixotropic. Thixotropy is a time dependent phenomenon brought about by the breakdown of the fluid structure during shearing. This results in a reduction in resistance to flow until an equilibrium level is reached (Fig. 2.10). After shearing has ceased the structure gradually reforms and thus thixotropy is reversible. Thixotropy aids wetting of the substrate during spreading, in adhesives which otherwise require a high viscosity.

Usable life – as soon as the resin and hardener components of a cold cure product are brought together they begin to react and the cross-linking process commences. The rate of cross-linking, and therefore hardening, depends on the reactivity of the formulation and the mobility of the molecules. After a time the mix becomes

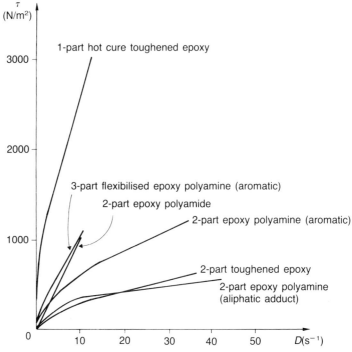

Fig. 2.8. Flow curves for a range of adhesives at 20 °C (Ref. 14).

stiff and unworkable and has come to the end of its usable or 'pot' life. Whereas pot life defines the limit of workability, gel time is taken to be the point at which solidification commences. In some circumstances the two values are very similar but pot life is the more meaningful. It may be assessed in a number of ways, e.g.

(a) the time taken for a specified quantity of material in a specified container to reach a given temperature(15)
(b) by measurement of the time at which joints made with the mixed material no longer give satisfactory performance(16)
(c) by assessing the time at which less than a specified area of filter paper is penetrated on wetting(17)
(d) by measuring the time to reach a viscosity above which the mixed material becomes unworkable, e.g. at 150 Pa s.(14)
(e) by monitoring the change in viscosity with time and identifying a change point at which the gradient of the curve rapidly increases(18).

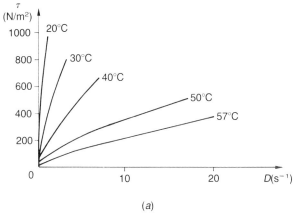

(a)

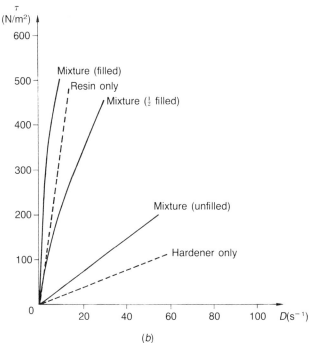

(b)

Fig. 2.9. Influence of temperature and filler content on flow curves (Ref. 14). (a) Temperature effect on 1-part hot cure toughened epoxy. (b) Filler content effect on 3-part flexibilised epoxy polyamine.

Table 2.4. *Pot life in minutes for a range of adhesives, determined by various methods (Ref. 14)*

Pot life definition (source)	2-part epoxy polyamide	2-part epoxy polyamine (aliphatic)	2-part epoxy polyamine (aromatic)	3-part flexibilised epoxy polyamine	2-part toughened epoxy
Temperature of 200 g insulated sample reaches 40 °C (Ref. 15)	100	50	24	60	66
Wetted area of filter paper falls below 10% (Ref. 17)	75	—	—	—	65
Viscosity at shear rate of 10 s^{-1} exceeds 150 Pa s (Ref. 14)	74	68	44	60	82
Change point in gradient of flow curve (Ref. 18)	98	48	40	39	60

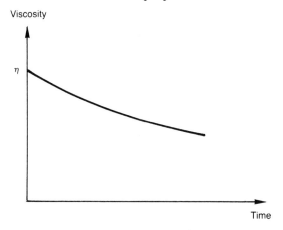

Fig. 2.10. Thixotropic fluid.

Table 2.4 shows how the pot life varies between several epoxy adhesives and according to the method of assessment.

Wetting ability. – The ability of an adhesive to wet a substrate surface (and the contact angle formed between liquid and solid) is fundamental to obtaining adhesion (Chapter 3). It is feasible to measure such contact angles for a stationary drop of liquid on a solid surface but the high filler contents of many structural adhesives mean that spreading is controlled by viscous forces. Adhesives being applied towards the end of their usable life tend to lack wetting ability.

Joint open time. – This starts when the adhesive has been applied to the parts to be joined and represents the time limit during which the joint should be closed, otherwise the strength of the bond may be reduced significantly. This is a function of the degree of hardening and reaction with the atmosphere itself. Open time may be reduced by high humidity and high temperature. The exposed surface of some hardeners e.g. aliphatic polyamines, can react with moisture and carbon dioxide in the atmosphere in a way which impairs adhesion between the two components to be joined. The effect can be minimised by lightly disturbing the surface of the wet adhesive immediately prior to closing the joint.

(3) During cure

Rate of strength development. The effect of temperature on curing rate will vary for different adhesives. In general, low temperatures increase the curing period considerably and many epoxy resin formulations stop curing altogether below 5 °C. A rule of thumb often quoted is that the curing period doubles for every 10 °C fall in temperature below ambient but halves for every 10 °C rise in temperature above ambient.

Fig. 2.11 illustrates how the manufacturer can influence the rate of strength development of a two-part epoxy by appropriate formulation(19). Compare these curves with the more rapid rates of gain of strength attainable with a structural polyester mortar and a general purpose cyanoacrylate 'super-glue'. For single-part products the rate of curing is controlled closely by the temperature applied. Fig. 2.12 shows a typical cure time/temperature relationship for a toughened epoxy(20).

(4) Hardened state

In general the properties of the adhesive in the hardened state are determined by its internal structure. Strength and elasticity derive

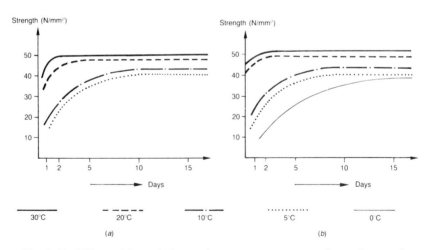

Fig. 2.11. Effect of formulation and cure temperature on flexural strength development of a two-part epoxy (Ref. 19). (*a*) Normal type. (*b*) Rapid type.

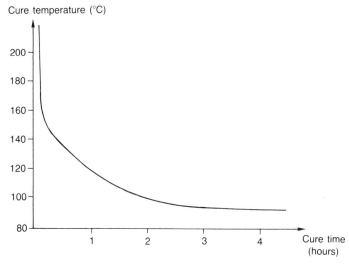

Fig. 2.12. Cure temperature/time relationship for a one-part epoxy (Ref. 20).

from molecular interactions, and a change in formulation to improve one property is likely to be at the expense of another. Further, the value measured is likely to be sensitive to the method of test, for example due to rate of loading or whether the adhesive is tested in bulk or thin-film form. Some important mechanical and physical properties of hardened adhesives are discussed below.

Strength and stress/strain characteristics. For satisfactory bonded joint design the important mechanical properties of the hardened adhesive under short-term loading are tensile and shear strengths, modulus of elasticity, elongation or strain capacity at failure, and fracture toughness (see Chapter 4). One initial experimental problem is the ability to fabricate reliable flaw-free bulk adhesive specimens with viscous cold-curing compounds. Work at Oxford Polytechnic(22) has demonstrated that the centrifuging of mixed adhesive can largely overcome such problems.

For the measurement of tensile properties, dumb-bell specimens of the form shown in Fig. 2.13 are suggested. Tensile modulus, Poisson's ratio and elongation at failure may be measured with appropriate strain monitoring equipment and a set of stress/strain curves for a typical range of epoxies is given in Fig. 2.14. Similar

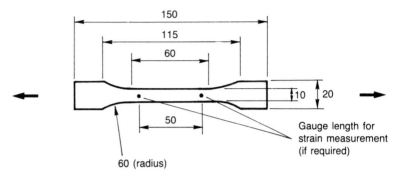

Specimen thickness = 3 mm

Fig. 2.13. Tensile dumb-bell specimen (Ref. 21).

specimens subjected to wet or moist environments may provide useful information on changes in strength or ductility due to water-induced plasticisation.

One way of monitoring the behaviour of a material in shear is with a torsional test. The method has the advantage that, provided a circular specimen is used, a condition of pure shear can be achieved. However, torsion tests can be relatively difficult to perform unless specimens can be machined accurately to avoid warping effects. The alternative is to test a prismatic specimen in a shear box of the form outlined in Fig. 2.15.

To assess flexural modulus of the hardened adhesive a specimen 200 × 25 × 12 mm deep tested in four point bending may be used. The sample under test is loaded transversely at the third points at a crosshead speed of 1 mm/minute and the central deflection recorded (Fig. 2.16). From the load–deflection curve, the secant modulus at 0.2% strain may be calculated. A lower limit on flexural modulus may be specified to prevent problems due to creep of the adhesive under sustained loads, whereas the upper limit will be to reduce stress concentrations arising from strain incompatibilities, for example at changes in section.

An alternative method of bulk property determination has been devised at Oxford Polytechnic(22). This involves monitoring the compression of pencil-like specimens from which a large amount of data may be derived (Young's modulus, Bulk modulus, Poisson's ratio and Glass Transition Temperature (T_g)).

56

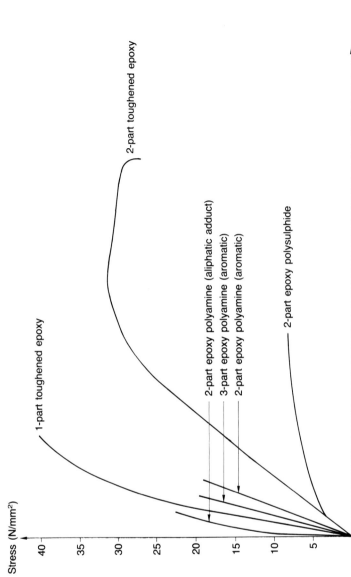

Fig. 2.14. Typical tensile stress/strain curves for a range of epoxy adhesives.

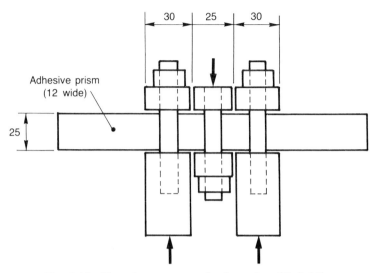

Fig. 2.15. Shear box test on adhesive prism (Ref. 21).

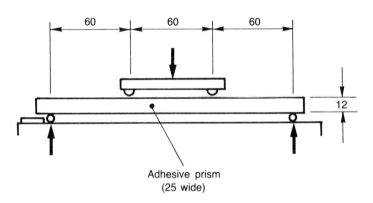

Fig. 2.16. Flexural test on adhesive prism (Ref. 21).

Table 2.5 summarises typical values of these bulk mechanical properties at around 20 °C for a range of epoxy adhesives. Of particular note is the low strength and stiffness of the epoxy polyamides and polysulphides as compared to those with aliphatic polyamine hardeners. For the design of bonded assemblies, joint tests are often used to determine the relevant mechanical properties. It must be remembered, however, that the results will be highly dependent on the specimen geometry and testing conditions and

Table 2.5. *Bulk mechanical properties of epoxies at ambient temperature*

Adhesive	Tensile strength MN m^{-2}	Shear strength MN m^{-2}	Tensile modulus GN m^{-2}	Flexural modulus GN m^{-2}	Poisson's ratio
2-part cold cure epoxy polyamide	12	15	2.0	2.1	0.40
2-part cold cure epoxy polyamine (aliphatic)	—	37	—	8.4	0.28
2-part cold cure epoxy polyamine (aliphatic adduct)	20	24	6.3	7.8	0.28
2-part cold cure epoxy polyamine (aromatic)	16	28	2.8	3.1	0.38
3-part cold cure flexibilised epoxy polyamine (aromatic)	—	25	—	5.1	—
2-part cold cure epoxy polysulphide	—	19	—	2.1	—
2-part cold cure toughened epoxy	—	25		1.9	

this aspect will be further discussed in Chapter 4. Nevertheless, joint tests can provide valuable comparative data. The stress–strain curves of Fig. 2.17 illustrate the variability in performance between different types of adhesives and between formulations of the same basic generic type as measured using tensile butt-joint tests(23).

Fracture toughness. The traditional approach to design using structural materials has been to compare the average stress or strain acting on the net cross-sectional area of the element with some ultimate stress or strain criteria. For the design of assemblies using brittle materials, which are sensitive to the presence of flaws, the theories of Fracture Mechanics have been introduced to overcome the shortcomings of the traditional approach. These recognise that the stress field around a crack or flaw can be defined uniquely by a parameter termed the stress intensity factor (K) which is directly proportional to the applied load. When K reaches some critical

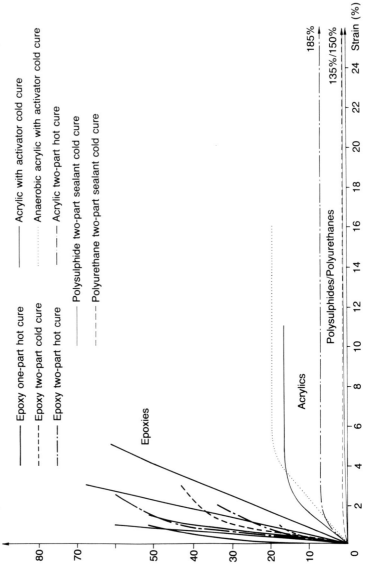

Fig. 2.17. Typical stress–strain curves for adhesives from tensile butt-joint tests (Ref. 23).

value, K_c – the fracture toughness of the material, a previously stationary or slow moving crack will jump ahead. Since some adhesives tend to be brittle in nature, especially at lower temperatures, the fracture toughness can be an important material design parameter. K_c is not unique but is dependent on factors such as the temperature of testing and the applied strain rate.

To measure bulk mode I fracture toughness, K_{IC}, (tensile load acting normally to the crack surface – Fig. 2.18(a)) single-edge notched (SEN) beam or tension specimens are recommended, as illustrated in Fig. 2.19. These specimens suffer two disadvantages, namely that failure is always catastrophic and the calculation of stress intensity factor requires that the crack length be known. Typical values of K_{IC} are summarised in Table 2.6(a) for a range of epoxy adhesives. Note particularly the increase in K_{IC} achieved by 'toughening'. Typical values of plain strain fracture toughness for other materials are given in Table 2.6(b) for comparison.

When considering fracture of the bulk adhesive, Mode II (shear) or Mode III (mixed) (Fig. 2.18(b) & (c)), fracture need not be considered since Mode I is the lower energy and therefore critical fracture mode, although in joints mixed mode loading may give rise to a lower critical fracture toughness. It must also be remembered that the fracture toughness of a joint may be controlled by the adhesive/substrate interface rather than that of the bulk adhesive and by the bondline thickness (see Chapter 4). Thus, for the design of bonded connections, measurement of the adhesive joint fracture toughness may be more appropriate.

Temperature resistance. Most synthetic adhesives are based on polymeric materials and as such exhibit properties which are characteristic of polymers. This is particularly so when considering their response to temperature variation. At a certain temperature, known as the Glass Transition Temperature (T_g), polymers change

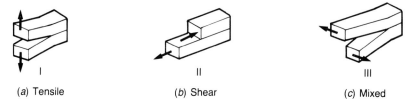

| (*a*) Tensile | (*b*) Shear | (*c*) Mixed |

Fig. 2.18. Principal fracture modes.

61

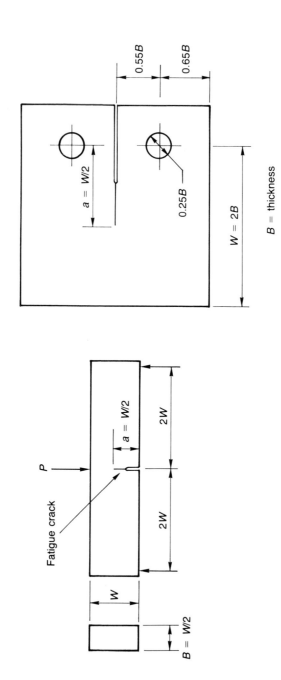

Fig. 2.19. Fracture toughness specimens. (*a*) Single-edge notched beam specimen. (*b*) compact tension specimen.

Table 2.6. *Mode I fracture toughness* K_{IC}

		K_{IC} (MPa m$^{\frac{1}{2}}$)
(a)	Epoxy adhesives	
	2-part cold cure epoxy polyamide	0.5
	2-part cold cure epoxy polyamine (aliphatic)	2.3
	2-part cold cure epoxy polyamine (aliphatic adduct)	1.6
	2-part cold cure epoxy polyamine (aromatic)	0.8
	3-part cold cure flexibilised epoxy polyamine (aromatic)	1.6
	2-part cold cure toughened epoxy	5.0
	1-part hot cure toughened epoxy	3.0
(b)	Other materials	
	steels	40–200
	aluminium alloys	23–66
	thermoplastics	1–2.2
	concrete	0.3–1.3
	glass	0.3–0.6
	timber (Douglas fir)	0.3

from relatively hard, elastic, glass-like substances to relatively viscous rubbery materials. The transition temperature will vary from one polymer to another. In addition, the transition temperature is dependent on the rate of loading if the measurement process involves mechanical deformation. The relationship between temperature and time effects will be discussed later in this Chapter.

Classical methods for measuring T_g include thermal, electrical, optical and dynamic mechanical techniques. For the civil engineer, quasi-static mechanical methods utilising a flexural test on a hardened adhesive prism are more convenient for determining the Heat Distortion Temperature (HDT). However, the results of such methods are dependent on the specimen configuration and the rate of loading selected and as such are only accurate for comparative purposes. A typical 'Heat Distortion Test' might utilise the same specimen configuration as for assessing flexural modulus (Fig. 2.16). The sample is placed in a temperature-controlled cabinet and a constant load is applied to achieve a maximum fibre stress of 1.81 N/mm^2 in accordance with BS 2782 (24). The central deflection at room temperature (say 20 °C) is then recorded. The HDT of the adhesive is taken as the temperature, measured on a thermocouple attached to the specimen, attained by the sample after undergoing a further 0.25 mm deflection while subject to a surface heating rate of 0.5 °C/minute. The HDTs of a range of cold cure epoxy adhesives measured in this manner are summarised in Table 2.7.

It will be noted that these HDT values lie in the range 34 to 48 °C which for many civil engineering applications may not be much in excess of anticipated maximum service temperatures. For example, on the soffits of concrete bridges temperature extremes in the UK may lie between −20 °C and +38 °C (25). In steel bridges maximum temperature extremes of 60 to 65 °C may occur locally and this is one reason why single part hot cure epoxy products which have higher T_g values of the order of 100 °C and more are preferred in such situations.

The influence of temperature on basic mechanical properties of the hardened adhesive such as bulk flexural modulus and shear strength is illustrated in Figs. 2.20 and 2.21, respectively (26). From these figures it is evident that the response of all five adhesives to temperature variations within the range 15 °C–65 °C is similar. The most noticeable feature of the curves is the rapid deterioration in both stiffness and strength at a temperature close to the measured HDT of the adhesive.

This general correlation between the HDT of the adhesives and the temperature range within which their basic engineering properties undergo significant change, confirms the proposition that the HDT is associated with a temperature at which important changes in molecular structure of the adhesive are occurring and which affect the way in which adhesives are capable of carrying load.

Table 2.7. *HDT of cold cure epoxies*

Adhesive	HDT (°C)
2-part cold cure epoxy polyamide	40
2-part cold cure epoxy polyamine (aliphatic)	41
2-part cold cure epoxy polyamine (aliphatic adduct)	43
2-part cold cure epoxy polyamine (aromatic)	48
2-part cold cure epoxy polysulphide	34

Adhesive properties

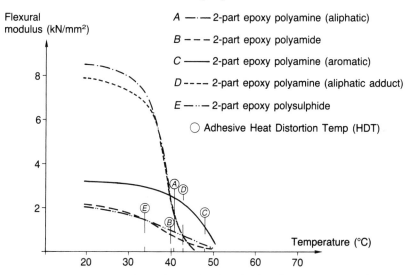

Flexural modulus (kN/mm²)

A — ·— 2-part epoxy polyamine (aliphatic)

B — — — 2-part epoxy polyamide

C ——— 2-part epoxy polyamine (aromatic)

D — — — — 2-part epoxy polyamine (aliphatic adduct)

E — ·· — 2-part epoxy polysulphide

○ Adhesive Heat Distortion Temp (HDT)

Fig. 2.20. Temperature dependence of bulk adhesive flexural modulus (Ref. 25).

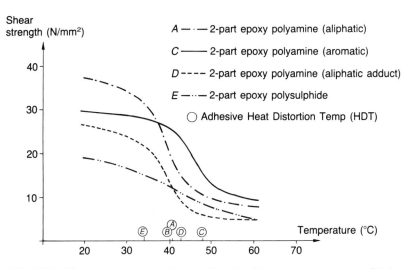

Shear strength (N/mm²)

A — ·— 2-part epoxy polyamine (aliphatic)

C ——— 2-part epoxy polyamine (aromatic)

D — — — — 2-part epoxy polyamine (aliphatic adduct)

E — ·· — 2-part epoxy polysulphide

○ Adhesive Heat Distortion Temp (HDT)

Fig. 2.21. Temperature dependence of bulk adhesive shear strength (Ref. 25).

65

One final point worth noting is that, provided the temperature does not rise above 150–200 °C, when a chemical deterioration occurs, the reduction in mechanical properties above the HDT is reversible on cooling. Indeed, stiffness and strength may even be enhanced due to post-cure effects at the elevated temperature. On the debit side, T_g or HDT will be lowered by water absorption into the polymer. Indeed, the T_g should be one of the first criteria in assessing the likely suitability of candidate adhesives.

Moisture resistance. All adhesives are susceptible, to some degree, to the effects of exposure to water or water vapour. Indeed, when cured, the very epoxy groups which give epoxies their adhesive properties also render them hydrophilic(27). This water uptake is accommodated largely by swelling. The effect of exposure to moisture is to alter the adhesive's properties, often in some undesirable way. Water may enter an adhesive either by diffusion or by capillary action through cracks and crazes. Once inside, the water may alter the properties of the adhesive either in a reversible manner, for example by plasticisation, or in an irreversible manner, for example by hydrolisation, cracking or crazing.

To determine water transport properties, thin film specimens may be immersed in water at a known temperature or stored in an atmosphere of known humidity and temperature. The water uptake of the adhesive is then measured and the fractional uptake plotted against the square root of time per unit thickness to comply with the mathematics of diffusion. Results for $60 \times 12 \times 2$ mm specimens of five cold-cure epoxies immersed in water at 20 °C are shown in Fig. 2.22. The specimen size was selected to enable flexural modulus tests to be conducted. If the plot shows a linear increase followed by an equilibrium plateau then the uptake is termed 'Fickian'. S-shaped plots are 'Non-Fickian' and are thought generally to be typical of glassy polymers. The relatively unmodified polyamide and the aliphatic polyamine absorb more than 5% by weight of water. However, in the latter case independent research (28) using electron microscopy revealed extensive micro-cracking in the resin and pronounced discontinuities at the surface of the relatively large (up to 0.4 mm) silica filler particles. Supplementary dye-penetration experiments with this adhesive confirmed that penetration was rapid in some areas. Conversely, the aromatic polyamine and the adducted, and therefore modified, aliphatic polyamine absorb less than 1% by weight. The polysulphide lies in the middle of the range. All five

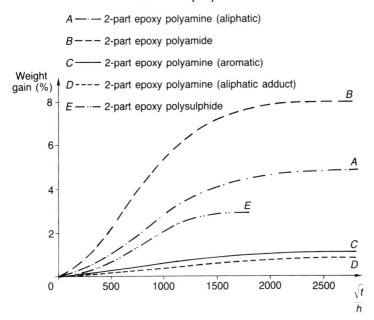

Fig. 2.22. Water uptake plots for a range of epoxy adhesives (Ref. 26).

adhesives are weakened by water absorption, as measured by bulk shear strength (Fig. 2.23). With the exception of the aromatic polyamine, this is associated with plasticisation of the adhesives, as measured by the flexural modulus of the specimens (Fig. 2.24) (26). However, fracture toughness was increased, at least in the short term.

With regard to adhesive joints, strength loss may be dictated by adhesive plasticisation as discussed above or by displacement of the adhesive from the substrate on water 'wicking' along the interface, or both. Such effects on joint properties will be considered in some detail in Chapter 4.

Creep. In general, polymers exhibit a degree of visco-elastic behaviour and thus for full characterisation of such a material a knowledge of its rate dependent response is necessary. To determine the long-term behaviour of a material either stress relaxation or creep tests may be used. The former involves monitoring the time-dependent change in stress which results from the application of a constant strain to a specimen at constant temperature. Conversely,

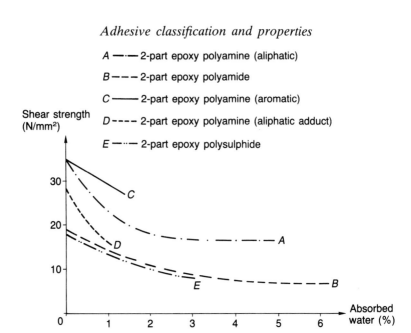

Fig. 2.23. Moisture dependence of bulk adhesive shear strength (Ref. 26).

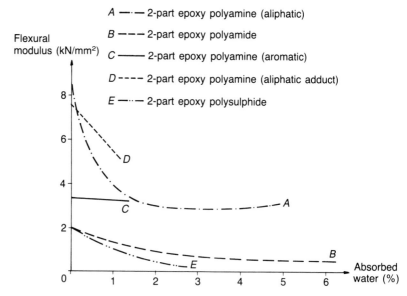

Fig. 2.24. Moisture dependence of bulk adhesive flexural modulus (Ref. 26).

creep can be thought of as time-dependent flow under constant load which may lead to fracture or creep rupture.

There are three main parameters affecting creep, namely stress, time and temperature. Moisture can also affect the creep of absorbent materials, such as some of the structural adhesives. During creep experiments the values of stress and temperature are kept constant. As different materials exhibit different creep properties, a method of characterising creep is required. This is usually in terms of its creep modulus (M_t) given by

$$M_t = \frac{\sigma_o}{\epsilon_t}$$

where

σ_o = constant applied stress

and

ϵ_t = total strain at time t.

Alternatively the creep compliance (C_t) can be obtained from

$$C_t = \frac{\epsilon_t}{\sigma_o}$$

A conventional creep curve as exhibited by most materials is illustrated in Fig. 2.25 although many engineers present the data using log axes to produce a graph of the form shown in Fig. 2.26. Data from families of strain–time curves at various values of constant stress are used to produce isochronous stress–strain curves (Fig. 2.27). These are obtained by cross-plotting stresses and strains at various times from the commencement of loading. The results of creep tests can also be used to derive constant strain, or isometric, curves of stress versus time, also as illustrated in Fig. 2.27.

Creep tests on structural adhesives can be divided into tests on bulk hardened adhesive specimens and tests on adhesively bonded joints. The former provides information on the mechanical properties of the adhesive rather than the joints made from them. Fig. 2.28 displays the change in creep modulus with time for a range of cold-cure epoxy adhesives(26). These curves were derived from four point bend tests on adhesive prisms loaded in accordance with Fig. 2.16 using extreme fibre stresses ranging from 0.25 to 2.0 N/mm². The curves represent the stability of the adhesive with time under

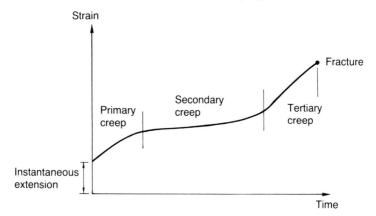

Fig. 2.25. Conventional creep curve exhibited by most materials.

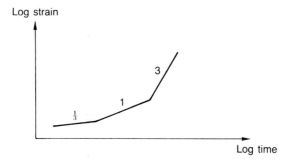

Fig. 2.26. Logarithmic creep curve.

sustained load and in this respect a flatter curve is beneficial. However, the relatively low long-term modulus values for the polyamide and polysulphide epoxies may give some cause for concern as to their potential structural efficiency under sustained load. In general, the more highly cross-linked the hardened adhesive structure and the higher the curing temperature, and hence T_g, the better the creep resistance.

Creep curves obtained from experiments using bulk hardened specimens do not necessarily compare with those obtained from joints under similar stress conditions due to the nature of adherend restraint. A further contributing factor to this difference is the reduction in stress concentrations which will occur in joints during creep.

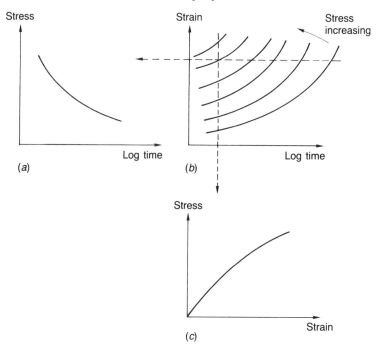

Fig. 2.27. Isometric and isochronous curves taken from a set of creep data. (*a*) Isometric stress v. log time. (*b*) Creep curves. (*c*) Isochronous stress v. strain.

Creep rate varies with stress level – generally the higher the stress, the greater the creep rate. It has also been suggested (29) that when the sustained stress is lower than some equilibrium value then indefinite creep will not occur. When stressed above this value the material will creep to failure. What happens to adhesive joints at the lower stress levels is perhaps more important in civil engineering than rapid creep to failure at high stress levels. This is so because structural adhesive joints tend to be designed to withstand low mean stresses, for example 10% of ultimate, but which have to be sustained for many years. Creep rate also varies with temperature. An increase above room temperature results in an increase in creep rate until the T_g is reached, when there is a marked further increase in creep rate.

It has already been noted that the T_g is sensitive to rate of loading and this has led to the development of time/temperature superposition techniques being used to characterise the response of polymer

71

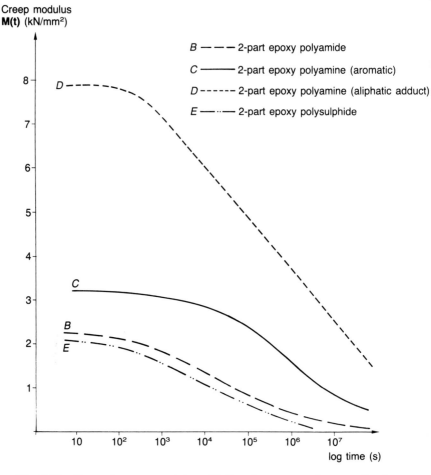

Fig. 2.28. Change in creep modulus with time for a range of epoxy adhesives (Ref. 26).

systems. If a family of property curves (e.g. strain v. time or log time) is plotted at a series of temperatures all the curves can be shifted parallel to the time axis until superposition produces a master curve at some reference temperature. This curve can then be used to predict the sample behaviour that would be obtained at the reference temperature if the property concerned was measured directly over the wide time scale which results. The technique permits the prediction of long-term creep behaviour at different temperatures and load conditions from limited short-term data.

The superposition approach can be used to produce a constitutive equation which expresses the creep compliance (C_a) of the adhesive in terms of a reference creep compliance (C_r) and shift factors for stress (a_σ), temperature (a_T) and resin content (a_v) such that $C_a = C_r \times a_\sigma \times a_t \times a_v \times t^m$. The method has been used by Dharmarajan et al. (30) to characterise the creep behaviour of epoxy, polyester and acrylic mortars in the form of prism specimens under 3 point loading. From relatively short-term tests, strain v. time curves such

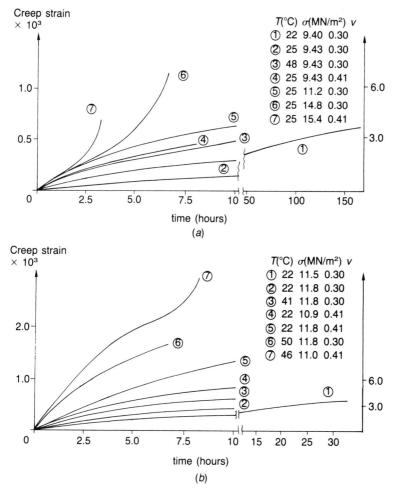

Fig. 2.29. Strain v. time curves for polyester and epoxy concretes (Ref. 30). (a) Polyester concrete. (b) Epoxy concrete.

73

as those shown in Fig. 2.29 have been used to produce reference curves as illustrated in Fig. 2.30. Data have shown the time exponent *m* to be independent of the polymer matrix but to have a value of 0.6 in flexural loading and about 0.2 in tension or compression. The transition from stable creep ($m < 1$) to creep rupture ($m > 1$) appears to occur at an equilibrium stress of between 45% and 55% of the short-term ultimate strength of the system.

Fatigue. The fatigue performance of an adhesive under cyclic loading will be influenced by the visco-elastic nature of the material and its resistance to crack propagation, or fracture toughness. At low frequencies and high temperatures, visco-elastic effects will predominate in a similar manner to that experienced with creep. At higher frequencies and lower temperatures fracture due to crack propagation either within the adhesive layer or at the adhesive/substrate interface will tend to control the number of load cycles that can be sustained prior to failure. Because the fatigue performance of an adhesive in a structural joint is closely linked to the joint

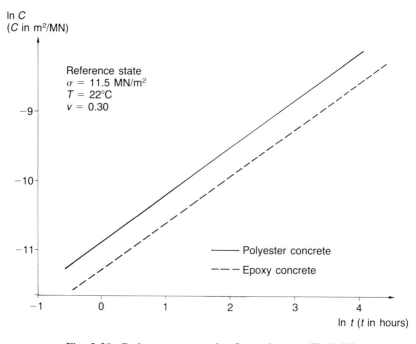

Fig. 2.30. Reference curves for flexural creep (Ref. 30).

74

configuration and stress distribution within the joint further discussion on fatigue will be deferred to Chapter 4.

2.4 Concluding remarks

Proper design of bonded joints requires an appreciation of the properties of the different materials to be joined as well as quantitative data on the properties of structural adhesives. In joint design, full account must be taken of the poor resistance of adhesives to peel and to cleavage forces; shear strength itself is unlikely to be a limiting factor. Therefore, a way forward might be to sacrifice some shear strength for high toughness and peel strength, such as that obtainable with flexible or toughened epoxies.

To limit the effects of creep under sustained load, an adhesive possessing a glass transition temperature, T_g, well above the service temperature is required. The use of such an adhesive should also result in the improved environmental durability of the bonded assembly.

75

CHAPTER THREE

Adhesion and surface pretreatment

3.1 Introduction

The successful performance of many every-day products, and many common materials and construction techniques, is dependent upon adequate adhesion between two or more constituents. Most engineers, however, have only the haziest of ideas about the whole concept of adhesion. For to know 'how' to prepare substrate surfaces for bonding does not necessarily require a knowledge of 'why' adhesive materials should stick to them. It is the intention of this chapter to connect theory with practice, to enable the reader to appreciate 'why' before discussing aspects of surface pretreatment pertinent to applications of adhesives in construction.

The strength of bonded assemblies depends not only on the cohesive strength of the adhesive, but also on the degree of adhesion to the bonding surface. One of the disadvantages of adhesive bonding as a method of fastening is that the surfaces need to be clean and, whatever their chemical nature, coherent if a satisfactory degree of contact, and therefore adhesion, is to be obtained. Frequently the adhesive itself is wrongly blamed for 'not sticking', but the general source of trouble lies with the surface pretreatment. The use of cold-cure epoxies generally necessitates the careful preparation of metallic adherends in particular, to ensure satisfactory long-term performance. Indeed, whatever the nature of the substrate, or the adhesive to be used with it, its pretreatment is probably the single most important aspect of the bonding operation. Inadequate surface pretreatment is usually the main cause of joints failing in service.

High initial bond strength is generally not as important as the bond durability, as dictated by the environmental stability of the adherend–adhesive interface. Surface pretreatments, whilst greatly affecting bond durability, generally have less effect on initial strength. Water is the substance which usually gives rise to problems in joint durability, with failure often being exhibited at, or near, the adherend–adhesive interface. The most important factor is the environmental stability of that interface, and appropriate surface pretreatment is viewed as the best means of maintaining adhesion under adverse conditions. Among several publications relevant to this crucial aspect of structural adhesive bonding are those by Comyn(1), Kinloch(2), Brockmann(3) and Hutchinson(4). The deleterious effects of water on joint strength, especially in combination with an applied tensile stress, may be appreciated with reference to Figs. 3.1 and 4.20.

The purpose of surface preparation is to remove contamination and weak surface layers, to change the substrate surface geometry, and/or introduce new chemical groups to provide, at least in the case of metals, an oxide layer more 'receptive' to the adhesive. An appreciation of the effects of pretreatments may be gained from surface analytical or mechanical test techniques. Experimental assessments of the effects of surface pretreatment, even when using appropriate mechanical tests, are of limited value unless environmental exposure is included. Self-stressed fracture mechanical cleavage specimens, as discussed in Chapter 4 and in the texts edited by Kinloch(2,5) for example, are therefore referred to wherever possible.

3.2 Interfacial contact and intrinsic adhesion

Preliminaries

The basic requirements for good adhesion are very simple, viz:

(1) intimate contact between adhesive and substrate
(2) absence of weak layers or contamination at the interface.

When two materials are bonded the resultant composite has several constituents and interfaces, as depicted for example in Fig. 3.2. Being liquid, adhesives flow over and into the surface irregularities of a solid, so coming into intimate contact with it and, as a result, interatomic forces are brought into play. Adhesives therefore join materials primarily by attaching to their surfaces within a layer of molecular dimensions, i.e. of the order of 0.1–0.5 nm. In joints involving metallic or siliceous substrates, the adhesive sticks to the surface oxide layer and not to the solid itself. In simple terms, there is an obvious conflict between having an adhesive material which

77

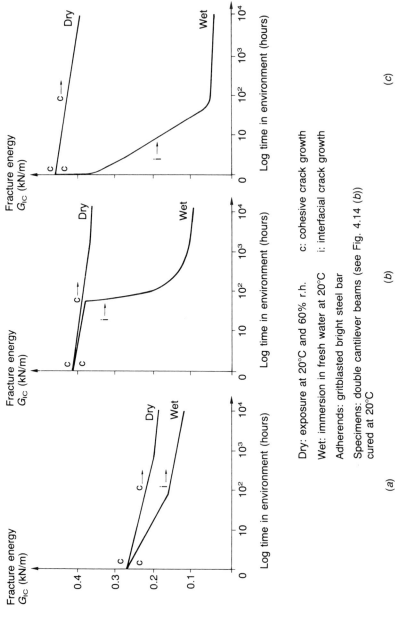

Fig. 3.1. Effect of water on the fracture energies of bonded joints, as a function of time. (a) Aromatic amine cured epoxy. (b) Aliphatic amine cured epoxy. (c) Epoxy polysulphide.

Dry: exposure at 20°C and 60% r.h.
Wet: immersion in fresh water at 20°C
Adherends: gritblasted bright steel bar
Specimens: double cantilever beams (see Fig. 4.14 (b)) cured at 20°C

c: cohesive crack growth
i: interfacial crack growth

(a)   (b)   (c)

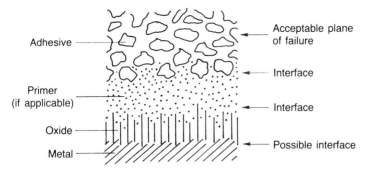

Fig. 3.2. Elements of a metal adherend/adhesive interface.

spreads and adheres well to the substrate, and one which when cured is a highly cross-linked structure possessing significant cohesive strength. Elevated temperature curing provides one solution to this dilemma since viscosity will initially be lowered, so facilitating the flow of the material over the surface together with increased molecular mobility. In cold-cure products the presence of mobile mono-amines greatly enhances the wetting potential.

Adhesion is often discussed in relation to the strength of joints, but the force required to fracture a joint is resisted by a complex interaction of internally generated bondline stresses. Attempts to use joint strengths as a measure of adhesion can therefore be extremely misleading. In the ensuing discussion, the term adhesion is reserved for bonding across interfaces, and there are many useful recent publications on the science of adhesion(6–15).

Interfacial contact

The adhesive has to spread over the adherend surface, penetrating its irregularities, displacing air and any contaminants present. The ideal conditions for this to be realised are that:

(1) the surface free energy of the adherend should be higher than that of the liquid adhesive
(2) the liquid adhesive should exhibit a zero or near zero contact angle
(3) the adhesive's viscosity should be relatively low at some time during the bonding operation

79

(4) the joint should be closed at a rate and in a manner that assists air displacement
(5) there should be an extended time before setting
(6) an external driving pressure should be applied. (It is sometimes suggested, for example in concrete repair work, that the 'adhesive' be scrubbed into the surface.)

Wetting equilibria and contact angles

Clearly the adhesive must wet the adherend, implying the common-sense idea of a thin film of liquid spreading uniformly without breaking into droplets on the surface (Fig. 3.3). The water break-free test is the simplest approach to a qualitative visual assessment; the thermodynamic approach to wetting allows quantitative study.

Surface tension is a direct measurement of intermolecular forces. At the surface of a liquid, there is an imbalance of attractive forces between neighbouring molecules, so that work has to be done on molecules to bring them to the surface. Surface molecules thus possess a higher energy than those within the bulk liquid and this extra energy is termed the 'surface free energy' or 'surface energy'. Surface energy and surface tension are dimensionally equivalent and numerically the same, and are represented by γ. Surface energy

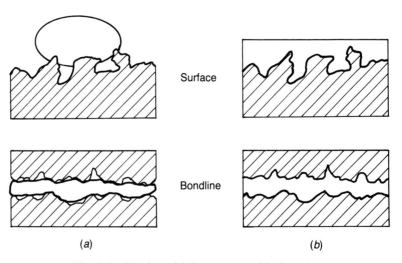

Fig. 3.3. Wetting. (*a*) Incomplete. (*b*) Complete.

80

may generally be expressed by two terms, namely a dispersion, γ^D, and a polar, γ^P, component, which may be determined experimentally, such that

$$\gamma = \gamma^D + \gamma^P$$

Fig. 3.4 illustrates the surface tension vectors at the three-phase point of contact of solid, liquid and vapour. The Young equation relating these tensions to θ is

$$\gamma_{SV} = \gamma_{SL} + \gamma_{LV} \cos \theta$$

γ_{SV} represents the surface free energy of the solid substrate resulting from adsorption of vapour from the liquid, and will be lower than the surface free energy of the clean bare solid surface in vacuo, γ_S. Dupré considered the work needed to separate a layer of liquid from a solid surface, viz

$$\text{Work} = \begin{pmatrix} \text{energy of new} \\ \text{surfaces created} \end{pmatrix} - \begin{pmatrix} \text{energy of} \\ \text{interface destroyed} \end{pmatrix}$$

This is the work of adhesion, W_A, and is given by

$$W_A = \gamma_S + \gamma_{LV} - \gamma_{SL}$$

The difference between γ_S and γ_{SV} is known as the spreading pressure, π_S, of the liquid's vapour on the solid surface. These two classic equations may then be combined so that

$$W_A = \gamma_{LV} (1 - \cos \theta) + \pi_S$$

Intuitively, optimum performance is expected if the liquid adhesive exhibits a zero or near zero contact angle, θ, giving a high value of W_A.

Real solid surfaces are not flat, so that observed contact angles

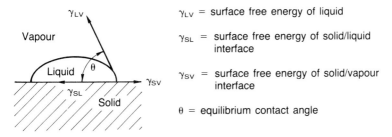

γ_{LV} = surface free energy of liquid

γ_{SL} = surface free energy of solid/liquid interface

γ_{SV} = surface free energy of solid/vapour interface

θ = equilibrium contact angle

Fig. 3.4. Liquid drop resting at equilibrium on a solid surface.

would be apparent only, and liquids may spread along fine pores and crevices by capillary action. Maximum wetting will be achieved when the capillary pressures are highest and viscosity is lowest, but surface micro-topography will introduce further complications(7,11, 16). The mixture of materials comprising most structural adhesives will influence wetting in a complex fashion. Selective adsorption of one component may provide dramatic changes in wetting rates or phase separation may occur, and both may be deliberately inbuilt into the adhesive formulation by the manufacturer to achieve a desirable balance of properties. Attempts to measure the contact angle of a drop of practical adhesive on a real surface are therefore fraught with considerable problems of experimental control and in subsequent interpretation. Surface wetting is, in fact, a kinetic phenomenon so that even if the ultimate equilibrium contact angle is zero, the advancing contact angle is never zero, but is a function of the rate of movement of the liquid and increases with increasing velocity(7,11). Hewlett and Pollard(17) devised a method for studying the dynamic contact angles of epoxies with regard to the repair of concrete with low viscosity injection resins. Unfortunately, they were unable to judge the various merits of the silane adhesion promoters incorporated in the resins.

Wetting involves a reduction in interfacial energy and Huntsberger(9) showed that, other than in exceptional circumstances, the free energy always decreases upon liquid/solid contact. It is concluded that, although most practical adhesives exhibit acute contact angles with their adherends, complete wetting may still occur so that the zero contact angle criterion for wetting (and therefore adhesive selection) is not valid. Bond performance may therefore be assumed to be determined by interfacial energies rather than by surface wetting.

Surface and interfacial free energies

Contact angles between solids and liquids of known surface tension can provide an indirect measurement of solid surface free energy, allowing a comparison between different surfaces. Just as liquids have surface tensions or surface energy, so do solid surfaces by virtue of the fact that they are surfaces. Surface tension of solids goes unnoticed because they are usually too rigid to be visibly distorted by the interatomic, rather than intermolecular, forces holding them together.* Zisman(18) introduced the useful distinction between high energy surfaces and low energy surfaces. Most liquids have surface free energies below 100 mJ m^{-2} with organic adhesives having low surface free energies – usually 50 mJ m^{-2}. Solid surfaces having similar free energies, such as plastics, are termed low energy surfaces. Hard solids, including the vast majority of metals and metal oxides, when atomically clean, have surface free energies typically in excess of 500 mJ m^{-2}; these are termed high energy surfaces. Some values for surface free energies of interest are collected in Table 3.1. An energetic surface will make a wetting

Table 3.1. *Values of surface free energies*

Surface	Suface free energy, γ, in vacuo (mJ m^{-2})	Young's modulus, E(GN m^{-2})	Material class
High energy surfaces			
C (diamond)	5140	1200	Diamond
Fe$_2$O$_3$ (ferric oxide or red haematite)	1357	210	
Al$_2$O$_3$ (alumina or aluminium oxide)	638	70	Metals
SiO$_2$ (silica or silicon dioxide)	287		
Low energy surfaces			
H$_2$O (water)†	72.2		
CFRP (heavily sanded)†	58		
Carbon fibre†	51.6		
Epoxide (amine-cured)	~42	4 (when cured)	Thermosets
Nylon	42		
PMMA (perspex)	40		
PVC	40		
Cyanoacrylate	~33		Thermoplastics
Polyethylene	33	2	
Silicones	25		Elastomers
PTFE	15.5		

†large polar component, γ^p

* The surface energy of a solid is roughly proportional to its Young's modulus. Hooke's law is really an approximation which arises from the character of the chemical bond between atoms. Thus the same interatomic forces give rise to both 'E' and to γ.

83

liquid spread on it, rather than remain as a discrete drop, so that adhesives should readily spread and wet the oxide layers of metallic substrates and the solid components of concrete (i.e. $\gamma_{SV} > \gamma_{LV}$).

The problem with high-energy adherend surfaces is that atmospheric contaminants are readily adsorbed on them, so reducing the surface free energy of attraction for the adhesive. Kinloch(2) suggests that the polar nature of structural adhesives will lead to displacement of the less polar, often hydrocarbon, contaminants. Polar water molecules, on the other hand, seem less likely to be readily displaced and, in view of their very small size in relation to an adhesive macromolecule, are able to adsorb in vast numbers on the surface.

Zisman(18) measured the critical surface tension, γ_C,* of a number of different metal and oxide surfaces which were exposed to atmospheres with controlled humidity. It was found that γ_C for all substrates was lowered to ~45 mJ m^{-2} at 0.6% r.h. and to ~37 mJ m^{-2} at 95% r.h. Gledhill *et al.*(19) extended this work to the wettability of mild steel substrates of different surface rugosities and deduced the value of γ_{SV} as a function of relative humidity (Fig. 3.5). Further, the enhanced wetting of gritblasted steel at low humidities was reflected directly in higher joint strengths. Hence

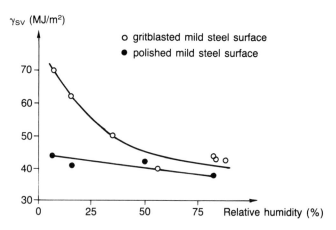

Fig. 3.5. Effect of humidity on surface free energy of steel (Ref. 19).

* γ_C is an empirical parameter, and is not the surface free energy of the solid. It equates to the surface tension of a liquid which just wets its surface with zero contact angle, and may be obtained graphically from the intercept of the plot of cos θ against γ_{LV} with the line cos θ = 1.

any clean hydrophilic surface such as metal, metal oxide or the exposed aggregate in concrete is converted, upon exposure to a humid atmosphere, from a high energy surface to a low energy surface, with a surface free energy barely greater than that of the adhesive. These considerations demonstrate the importance of conducting the bonding operation in an environment which is as clean and dry as possible. Similarly by priming a substrate the surface free energy is reduced considerably so that the primer/adhesive system must be chemically compatible.

Mittal(12) reports that a direct relationship between W_A or γ_C and joint strength has been found to exist for some adhesive/substrate systems. The interfacial free energy is the most important surface property in that the lower the value of γ_{SL}, the greater the theoretical adhesion. However, for the same value of γ_{SL} different adhesives yield different experimentally measured bond strengths, leading to the conclusion that joint strength or 'practical adhesion' cannot be equated directly with the thermodynamic work of adhesion, W_A. Mittal states that this discrepancy is because (a) during joint rupture a number of inelastic deformations occur with consequent dissipation of energy, and (b) W_A refers to a defect-free interface, which is never the case. Other authors(2,20,21) have shown that the thermodynamic work of adhesion in water is negative for typical metal oxide/epoxy interfaces, indicating that water is capable of displacing adhesive from the substrate if no other forces are involved. Thus bonds to high-energy polar adherends are theoretically unstable in the presence of water, and this is borne out by experience.

Mechanisms of adhesion

Once interfacial contact between the adhesive and adherend has been established under favourable thermodynamic conditions, adhesive curing enables stress to be transmitted. There is some controversy regarding the basic nature of the intrinsic forces then acting across interfaces which prevent them from separating under an applied load. The four main theories of adhesion which have been proposed are: mechanical interlocking, adsorption, diffusion and electrostatic attraction. Detailed treatments of these attraction theories can be found in the literature(6–15). The adsorption mechanism is generally favoured, with mechanical keying also playing an important role.

Mechanical interlocking. This theory proposes that mechanical keying of the adhesive into the irregularities or pores of the substrate surface occurs, and it underlies the layman's instinctive procedure of roughening surfaces to improve adhesion. Additionally, a rough surface will have a larger potential bonding area than a smooth one. As good adhesion to smooth surfaces such as glass is attainable, this theory seems limited to porous and fibrous materials such as textiles. Accordingly, penetration of adhesive into the micro-structure or porosity of high alloy metal oxide layers appears important. Thus micro-roughness, rather than macro-roughness, and consequent micro-mechanical inter-locking at a molecular scale is desirable with some metal surfaces, particularly in aiding the retention of adhesion under adverse conditions(2,3). In general, any improvement in joint strength from greater adherend rugosity may be ascribed to other factors such as the increased surface area, improved wetting, or enhanced energy dissipation of the adhesive during joint fracture(7,16).

Adsorption. This,˙ the most generally accepted theory outside Russia, proposes that with sufficiently intimate contact, the adhesive macro-molecules are physically adsorbed on to the substrate surface because of the forces acting between the atoms in the two surfaces. In effect, the polar nature of the adhesive molecules acts like a weak magnet and they are attracted towards polar adherend surfaces. The most common interfacial forces are van der Waals' forces, referred to as secondary bonds, although hydrogen bonding and primary chemical bonding (ionic or covalent) are involved in some cases. The terms primary and secondary imply the relative strength of the bonds (Table 3.2).

Primary bonding, although theoretically unnecessary to account for high joint strengths, may often increase measured joint strengths and is certainly of benefit in securing environmentally stable interfaces. Evidence of chemical bonding is generally limited to coupling agents, and Gettings and Kinloch(23,24) have provided strong evidence for chemical bonding at a particular silane primer/-steel interface.

The huge discrepancy between the magnitude of the attractive forces available for adhesion and measured joint strengths is attributed to insufficient interfacial contact, air voids, cracks, defects and stress concentrations.

Table 3.2. *Interactions which may contribute to adhesion (Refs. 2, 22)*

Type	Distance (nm)	Bond energy (kJ mol⁻¹)	Theoretical adhesion force (MN m⁻²)
Ionic	0.1–0.2	600–1000	
Covalent	0.1–0.2	60–700	17 500
Metallic	0.1–0.2	110–350	5,000
Hydrogen bonds	0.3–0.5	10–50	500
Dispersion (London) forces	0.3–0.5	0.1–40	60–360
Experimentally measured bond strength			<50

Diffusion. This theory proposes that adhesive macromolecules diffuse into the substrate, thereby eliminating the interface, and so can only apply to compatible polymeric substrates. It requires that the chain segments of the polymers possess sufficient mobility and are mutually soluble. The solvent welding of thermoplastics such as PVC (polyvinyl chloride), softened with a chlorinated solvent, is an example of such conditions being met. Diffusion will also take place when two pieces of the same plastic are heat-sealed. The joining of plastic service pipes for carrying gas and water makes use of the diffusion mechanism.

Electrostatic. The electrostatic theory, like the diffusion theory, originated in Russia. It is postulated that adhesion is due to the balance of electrostatic forces arising from the transfer of electrons between adhesive and substrate, resulting in the formation of a double layer of electrical charge at the interface. The two layers thus formed can be likened to the plates of a capacitor and work is expended in separating the two charged capacitor plates. Supportive evidence for this theory includes the fact that the parts of a ruptured joint are sometimes charged, but of course the charge may have been produced during failure. If, as claimed, the work of peeling polymeric films from metal surfaces is much greater than can be accounted for by van der Waals' forces, this may be due to the dissipation of energy through viscous and visco-elastic responses of the materials. The literature cites only a few special circumstances where the electrostatic mechanism may contribute to intrinsic adhesion.

3.3 Surface pretreatment

Opening remarks

Surface pretreatment involves:

(a) cleaning
(b) removal of weak surface layers
(c) re-cleaning.

The exact procedures required depend upon the nature of the adherends, and an indication of their inherent suitability for bonding is shown in Table 3.3. (The ranking order might be anticipated from the list of material surface energies collected in Table 3.1.) The first and main purpose of pretreatment is to clean the substrate, to remove contaminants and surface detritus (adsorbed principally through the van der Waals' forces referred to earlier). The next stage is to remove weak surface layers, usually by abrasion or etching, to provide a stable and coherent surface 'receptive' to the adhesive. The nature of the surface is directly affected by such treatments so that if, for example, the chemistry or topography of the surface is changed, its interaction with the adhesive will vary. The final (and commonly overlooked) stage is to re-clean the prepared surface to remove contamination, introduced by the various treatments, such as oil-mist, moisture, dust and chemical residues. Experimental evidence, however, suggests that solvent degreasing by brushing is not recommended.

Together with the literature detailing practical information on pretreatment procedures(25–33), there exist many research publications concerned with the optimisation or development of methods directed towards particular adherends. A visual appreciation of the effects of pretreatment has been facilitated in particular by electron microscopy in its various forms (e.g. Figs 3.6 and 3.7).

Safety precautions should be observed strictly where chemical solutions and solvents are employed for pretreatment procedures.

Methods of surface pretreatment

Some effects of surface pretreatments are summarised in Table 3.4.

Degreasing. Solvent degreasing removes grease and most potential contaminants. The choice of solvent should be based on the principle

Surface pretreatment

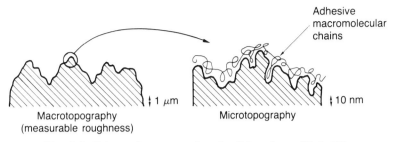

Fig. 3.6. Schematic topography of solid surfaces (Ref. 62).

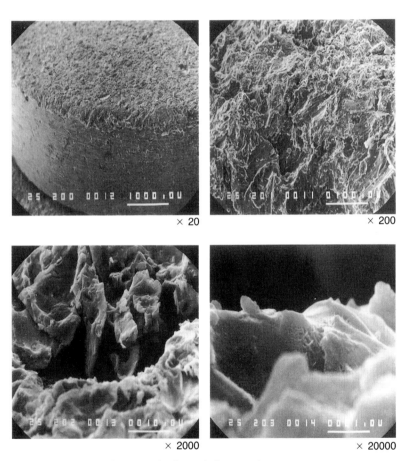

× 20   × 200

× 2000   × 20000

Micron scale shown in bottom r.h. corners

Fig. 3.7. Scanning electron micrographs of a gritblasted bright steel bar (surface inclined at 45°). Micron scale shown in bottom r.h. corners.

89

Table 3.3. *Pretreatment requirements*

Material	Suitability for bonding	Pretreatment required
Cast iron	*****	cursory
Steel	****	straightforward
Stainless steel	***	
Zinc	***	quite demanding
Aluminium	***	
Concrete	****	
GRP	****	straightforward
CFRP	***	
PVC	**	rigorous
Polyolefin	*	complex

Table 3.4. *Effects of surface pretreatments*

Treatment	Possible effects on surface	Substrate
Solvent etch	remove weak boundary layer	M, P
	weaken surface region by plasticisation	P
	increase surface roughness	P
Mechanical	remove weak boundary layer	M, P, C
	increase surface roughness	M, P, C
Chemical	remove weak boundary layer	M, P, C
	increase, or decrease, surface roughness	M, P, C
	alter surface chemistry with consequent changes in the rate and degree of wetting	M, P
Physical (e.g. flame, plasma, corona discharge)	remove weak boundary layer	M, P, C
	weaken surface region by plasticisation	P
	alter surface chemistry	M, P

Substrate abbreviations: M, metal; P, plastic; C, concrete.

that like dissolves like, although toxicity, flammability and cost should be taken into consideration.

Shields (29) counsels against re-using solvent since oils and greases may simply be redistributed thereby, and a monolayer of oil molecules suffices to constitute a weak boundary layer. Dipping in the solvent is unlikely to be successful, and the simplest form generally employed is brushing; ultrasonic or vapour degreasing systems are the most efficient. The latter enclosed systems are generally employed in mass production, in which the adherends to be degreased are suspended in a cooled zone over boiling solvent; vapour condenses on the cooled adherends and drops back into the bath so that only clean redistilled solvent makes contact. A volatile solvent such as acetone or trichloroethylene ('trike') should always be chosen, or else it may itself form a weak boundary layer; in other than fully enclosed systems 'trike' should be replaced by 1,1, 1-trichloroethane, which is much less toxic though more expensive.

For metallic substrates, alkaline cleaners and/or detergent solutions are often advised after solvent treatments, to remove dirt and inorganic solids. They may also be used instead of solvents, to obviate the safety problems such as flammability and toxicity, and should be followed by thorough rinsing and drying (in hot air) before bonding. The solvent etching of a polymer may have a similar roughening effect to the mechanical abrasion of metals(10), although the possibility that the surface regions of a polymer may be weakened by plasticisation should not be ignored.

Mechanical. Mechanical treatments often cause much obvious roughening of a surface but the effect on adhesion is complex(7,11, 16,34) and joints fabricated with highly polished mild steel adherends have shown increased strength and durability over gritblasted adherends for example. However, the real area available for adhesion is increased, the surface free energy level should be higher (because of the number of neighbourless atoms present at asperities), and the irregular profile should divert any propagating cracks into the bulk polymer. Against these potential advantages must be balanced the facts that, (a) many local interfacial stress concentrations are created, (b) proper wetting (especially by very viscous adhesives into deep narrow pits) may be rather difficult to ensure with consequent entrapment of air, and (c) segregation of the adhesive may occur with deleterious effect. If the joint is subsequently heat-cured, air trapped at the interface may rise into the bulk polymer, or else be compressed completely by the pressure of bonding in, say, an autoclave.

The various mechanical methods depend on the abrasive action of wire brushes, sand and emery papers, abrasive pads (e.g.

'Scotchbrite'), needle guns, or shot-blasting techniques to remove unwanted surface layers; these methods are generally more difficult to control than chemical methods.

Chemical. Chemical and electrochemical treatments tend to cause more complex changes than mechanical methods. In addition to cleaning action and the removal of weak layers, chemical treatments often roughen the surface microscopically. Anodising, for example, results in a very porous surface, and other techniques for metals result in a microfibrous topography(35). Satisfactory treatments for metals must result in the formation of stable and coherent oxide, and conditions such as the duration and temperature of the process may be critical. The changes in surface geometry and chemistry will affect the rate and degree of wetting. The surface chemical change will also alter the extent of interaction between adhesive and substrate, and may produce a chemically resistant surface layer which promotes bond strength retention under adverse conditions.

A significant disadvantage of chemical methods is the toxicity of the materials used, with a subsequent waste disposal problem.

Physical. Methods such as ionic bombardment (corona discharge) have proved successful with inert plastics, which are otherwise difficult to bond effectively because of their low surface energy (Table 3.1). It is probable that a local chemical conversion takes place on the polymer surface by oxidation, imparting a layer of higher surface free energy and thereby improving wettability(36). Allen *et al.*(37) report that corona discharge works quite well on aluminium and titanium surfaces, and ascribe this to the superior cleaning action, and therefore enhanced wetting, over solvent wiping. Initial lap joint strengths were, however, a little lower than obtained by employing chemically treated adherends.

Pretreatments for metals

Metal and oxide surfaces. In joints involving metallic substrates, the adhesive 'sticks' to the metal surface oxide layer and not to the metal itself. Such joints can cause problems in service because oxide structures, and bonds to them, are susceptible to environmental attack.

Refined metals, by their nature, are chemically unstable, tending to react with their environments and reverting on their surface to metallic compounds probably very similar to the minerals from which they were originally extracted. The most common such reaction is with oxygen, giving rise to a surface oxide layer, and the rate of reaction increases with increasing temperature. The higher the temperature, the thicker the oxide layer, and the more difficult it is to obtain satisfactory wetting by an 'adhesive', be it molten metal or an organic polymer. Because of the very high temperatures involved in fusion welding, the rapid surface oxidation necessitates very strict control of the process to achieve adequate bond or fusion. Hence most high alloy metals which oxidise rapidly, such as stainless steel and aluminium alloy,are particularly awkward to weld or to bond.

Environmentally stable metals with protective surface oxide layers are generally unsuitable for bonding without pretreatment because their oxide structures are mechanically weak and up to 3000 nm thick. These must be replaced by stronger, coherent and stable oxide structures. Conversely, iron and plain carbon steel require little surface treatment, provided their surfaces are free from rust and millscale, because the oxide layer (Fe_2O_3) is only about 3 nm thick. Under normal ambient conditions, the outermost surface oxygen groups hydrate, albeit much more slowly than the surface oxidises, to form a high density of hydroxyl groups. This surface then adsorbs several molecular layers of bound water which, for metallic and siliceous surfaces, are retained up to about 400 °C. It is these hydrated polar groups which form bonds with the polar organic resins(14,38), the relevant adhesion forces being dispersion and hydrogen bonds.

Steel. Publications specifically about steel surface preparation are by Sykes(39), Haigh(40) and Brockmann(41). Abrasive treatment is on the whole best for preparing plain-carbon steel, and any obvious rust or millscale should first be removed by wire brushing followed by degreasing. Sykes cautions that surfaces rusted in industrial or marine atmospheres will be contaminated by ferrous sulphate or chloride, and crystals of these salts are hard to dislodge even with blasting. Further treatment involving simple abrasion with emery cloth or wire brushes merely scores the surface, tending to rub contamination into the grooves which then becomes hard to dislodge; there is insufficient surface cutting action to encourage wetting by cold-cure adhesives. Brockmann(41), for instance, reports on the influence of the steel surface condition on which a one-part epoxy was heat-cured. By subsequently etching away the metal, then only in the case of shotblasted steel was a true replica of the steel surface morphology observed; the adhesive was found to have a porous texture when cured against ground- or degreased-only surfaces, indicating non-wetting. Figure 3.8 illustrates the very poor performance of emery abraded surfaces, with the specimen joints splitting apart completely along the adhesive–adherend interface within a matter of hours. Silane priming of the substrate surfaces was found to delay the time to failure, being particularly beneficial in combination with an epoxy polysulphide adhesive (see Fig. 3.14).

Shotblasting procedures, commonly employing angular chilled iron grit in the construction industry, remove inactive oxide and hydroxide layers by cutting and deformation of the base material leading to a fissured surface topography (Fig. 3.7). For other metals, alumina and carborundum are preferable hard, sharp abrasives. Shields(29) warns against the use of glass or metal beads of round shape as leading to peening of the surface. The size and nature of the abrasive grit should be matched to each type of metal and alloy, and different metal grits will leave traces of different metals on the blasted surface. Further, the geometry of the blasting nozzle and the pressure of blasting will affect the resultant topography. It is important to degrease the surface before abrasion, and to ensure that the grit particles are themselves free from contamination. Abrasive dust should then ideally be removed from the surface prior to bonding (a vacuum head fitted with edge brushes is recommended for structural steelwork(40)). Solvent degreasing, unless closed vapour systems are employed, is inadvisable because any contamination may simply be redistributed; there may also be other subtle undesirable surface effects. Figure 3.9 illustrates the deleterious effect of solvent degreasing following blasting. The use of wet blasting is limited in its applicability to non-corrosive metals unless anti-corrosion additives are included in the water; stainless steels can be blasted wet or dry with non-ferrous particles such as alumina, garnet or silica. Assessments of the surface finish of blast-cleaned steel for painting are covered by BS 4232(43) and SIS 05 59 00(44).

Chemical treatment of ferrous alloys with sulphuric or hydrofluoric acid, is not straightforward because of the precipitation of free carbon on the surface known as 'smutting', and the consequent need for 'de-smutting' immediately after etching(45). The metal then needs to be rinsed and washed in running water before transference

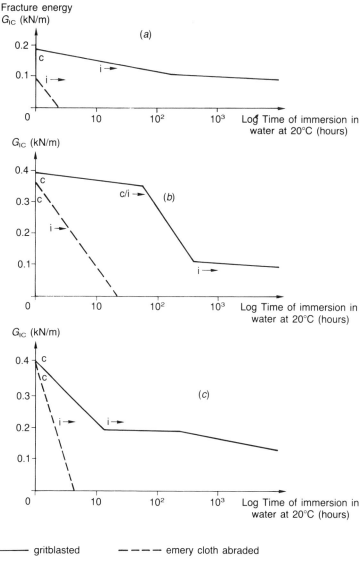

Fracture energy
G_{IC} (kN/m)

(a)

Log Time of immersion in
water at 20°C (hours)

G_{IC} (kN/m)

(b)

Log Time of immersion in
water at 20°C (hours)

G_{IC} (kN/m)

(c)

Log Time of immersion in
water at 20°C (hours)

——— gritblasted   — — — — emery cloth abraded

Specimens: steel wedge cleavage (see Fig. 4.14(c)) cured at 20°C

c: cohesive crack growth
i: interfacial crack growth

Fig. 3.8. Influence of surface abrasion on the fracture energies of bonded joints as a function of time of water immersion: gritblasting versus emery-cloth abrasion. (*a*) Aromatic amine cured epoxy. (*b*) Aliphatic amine cured epoxy. (*c*) Epoxy polysulphide.

95

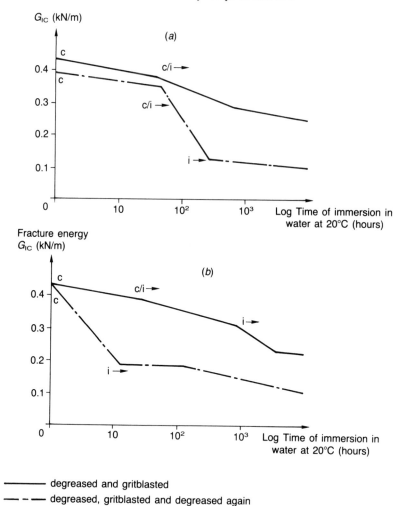

Fig. 3.9. Effect of solvent degreasing after gritblasting on the fracture energies of bonded joints as a function of time of water immersion (Ref. 42). (*a*) Aliphatic amine cured epoxy. (*b*) Epoxy polysulphide.

96

to a bath of isopropanol. It is, therefore, impossible thoroughly to clean and etch mild steel and to finish with a washing process without corrosion occurring. The temperature and duration of each of the operations are crucial to the resultant surface morphology, and to complete 'de-smutting'. Priming is advised as soon as possible after the final washing procedure. Trawinski(46) reports on a simple chemical process for plain carbon steel employing a nitric–phosphoric acid etchant, which is suitable for use at 21–27°C, and at remote locations. It is claimed to produce a smut-free microscopically rough surface, similar to that obtained with high temperature phosphoric acid-based etchants but with much safer resultant chemical residues. Further, superior performance was found in wedge cleavage tests over gritblasted and the usual phosphoric acid etch surface treatments, with a heat-cured epoxy and with or without primers.

The general term stainless steel is used not only for iron–nickel–chromium alloys of particular compositions, but often for any high alloy steels which are able to resist rusting in the presence of oxygen. The distinction between these classes is based on the types of (chromic) oxide film which form on their surfaces, and which retards the diffusion of aggresive ions. Oxygen (or an oxidising agent) is needed both to create this passive layer and to repair it in service; without it, stainless steel actively corrodes. Unfortunately, there is little published information on the effect of metal composition either on adhesive bond strength or interaction with etchant solutions. An acid etch pretreatment followed by 'de-smutting' and then priming is recommended: the number of etching procedures employed industrially are many and various. Gettings and Kinloch(24) found that the physical and chemical characteristics of the surface were strongly dependent on the manufacturing path of the steel, and that the environmental resistance of heat-cured epoxy joints was influenced both by the surface chemistry and the micro-roughness induced by chemical treatment.

Galvanised surfaces are characterised by relatively weak zinc oxide which may need to be removed by mechanical or, preferably, chemical means; a number of electrodeposition(35) and etch processes are described(47). Lees(48) cautions that although successful pre-treatment is possible, the zinc layer itself remains as a potential source of weakness, since it might be stripped off by a good adhesive. Adams and Wake(45) further note that certain adhesives and anhydride-cured epoxies may form soaps at the zinc–adhesive interface. Nevertheless, for only moderately demanding situations simple abrasion techniques may suffice, depending upon the nature of the galvanizing processes and the resultant surface.

Some comparisons of the effect of surface pretreatment on mechanical joint strength, or measurable adhesion, are reproduced in Tables 3.5–3.7. It is stressed again that the important consideration is the effect on long-term bond integrity, and not on short-term strength.

Aluminium. Since the Second World War, aluminium and its alloys have been extensively investigated in connection with aircraft industry requirements, namely for joints of the highest strength and greatest durability. Mechanical treatment and alkaline cleaning on their own result in poor durability; chemical etching and acid anodising are favoured, and rigorous pretreatment processes are documented for deoxidising and allowing a controlled re-oxidation of the substrate surfaces with powerful oxidising agents. The gradual replacement of chemical etching by anodising follows the discovery that the nature and thickness of the re-formed oxide surface depends as much on the wash which follows as on the etch itself(45).

In addition to the brief description of pretreatment methods in the general texts(26,29–32) are many specific publications(3,50–57). Contemporary opinion suggests that the function of pretreatment should be to produce a thick porous coherent oxide honeycomb, which imparts a degree of micro-mechanical interlock with the adhesive, and which is resistant to hydration. Hydration resistance and micro-mechanical interlocking have been the subjects of intensive study in recent years. A chromic–sulphuric acid etch has been favoured in Europe, whilst in the USA the Forest Products Laboratory (FPL) etch, based on sulphuric acid/sodium dichromate, has been optimised. The FPL etch is used alone or is followed by the Phosphoric Acid Anodising (PAA) process developed by Boeing. It has gradually been found that whilst etched adherends give higher initial bond strengths than do anodised ones, the latter confer more durable bonds. An explanation for this is that the oxide layer formed on etched adherends thickens with new, less coherent, oxide in the presence of moisture reaching the interface. Brockmann(3) suggests that the conventional phenolic resins confer an acid environment on the oxide surfaces, and it is this acidity which is inherently water-stable. Schematic sections through typical oxide morphologies are depicted in Fig. 3.10.

The choice of treatments for aluminium and its alloys revolves

Table 3.5. *Effect of pretreatment on lap shear strength of steel/polyvinyl-formal-phenolic adhesive joints*

	Substrate		
	Martensitic	Austenitic	
	stainless	stainless	Mild
	steel	steel	steel
Surface pretreatment	Mean bond strength $(MN\ m^{-2})$		
(A) Grit blast with No. 40 chilled iron shot.	35.4	28.4	30.2
(B) Vapour blast with garnet grit (200 grade and 400 grade) protect with oil, vapour degrease with trichloroethylene.	42.6	34.2	33.2
(C) Vapour degrease, immerse for 15 min at 65 °C in 5 pbw sodium metasilicate, 9 bpw 'Empilan NP4' detergent (Marchon Products), 236 pbw water. Rinse in hot distilled water, dry at 70 °C.	30.2	24.6	31.4
(D) Clean in proprietary alkaline solution.	35.6	22.2	25.9
(E) Etch 15 min at 50 °C in 0.33 pbw satd. sodium dichromate soln., 10 pbw conc. H_2SO_4. Brush off 'carbon' residue, rinse, dry at 70 °C.	40.0	14.88	28.2
(F) Vapour blast and etch (B and E).	42.8		
(G) Two-stage etch: 10 min at 65 °C in 100 pbw conc. HCl, 20 pbw formalin solution, 4 pbw 30% hydrogen peroxide, 90 pbw water. Rinse then etch for 10 min at 65 °C in 100 pbw conc. H_2SO_4 10 pbw sodium dichromate, 30 pbw water. Rinse and dry at 70 °C.	50.5	25.6	15.98
(H) Anodic etch for 90 s at 6 V in 300 g litre^{-1} H_2SO_4, rinse, dry at 70 °C.	45.4	24.8	39.8
(I) Anodic etch (H), passivate in chromic acid.	46.6	26.3	38.0
(J) Etch in 10% HCl (w/v) (5 min at 50 °C). rinse in 1% H_3PO_4, dry at 70 °C.	25.6	0.66	2.14
(K) Etch 10 min in 10% HNO_3 2% HF, rinse, dry at 70 °C.	45.5	22.2	28.0
(L) Etch (K), passivate in chromic acid.	46.4	23.8	31.2

Source: After Refs. 39, 49 and reproduced by kind permission of Elsevier Applied Science Publishers Ltd.

Table 3.6. *Effect of surface treatment of stainless steel on bond strength (EN58B steel-AV1566GB one component heat-cured epoxy (Ciba-Giegy))*

| | Lap shear strength (MN m^{-2}) | | | |
| | Initial | | After 30 days: water immersion at 40 °C | |
Surface treatment	23 °C	80 °C	23 °C	80 °C
Degrease in trichloroethylene.	20.9	20.0	14.7	17.5
Degrease, light grit blast (alumina grit), degrease.	24.8	31.4	16.0	18.3
Degrease, heavy grit blast, degrease.	26.3	28.6	13.2	16.2
Etch in 100 g litre^{-1} sulphuric acid, 100 g litre^{-1} oxalic acid, (15 min, 90 °C), desmut by brushing.	26.2	28.9	15.1	21.7
Etch in sulphuric acid/oxalic acid as above, desmut in sulphuric acid/chromic acid.	27.3	33.9	21.7	28.8

Source: After Refs. 39, 49 and reproduced by kind permission of Elsevier Applied Science Publishers Ltd.

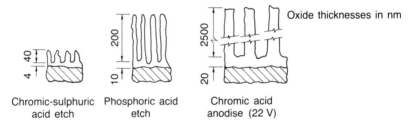

Fig. 3.10. Oxide morphology on aluminium-alloy after pretreatment (Ref. 58).

around the scale of operations, the metal composition, the adhesive to be used, the required durability, and the cost; thus there is a need for multiple recommendations. For example, for the potentially large-scale usage of aluminium in motor vehicle assembly Alcan International have developed an anodising process which is applied at the metal coil stage; the sheet material is then coated with lubricant for coil storage prior to pressing and then bonding.

Table 3.7. *Effects of pretreatment for steel on bond strength and water resistance (AV1566GB one-component heat-cured epoxy (Ciba-Geigy))*

		Lap shear strength (MN m^{-2})			
		Initial		After 30 days: water immersion at 40 °C	
Material		23 °C	80 °C	23 °C	80 °C
EN58B stainless steel	degrease	23.9	25.6	15.7	17.7
	grit blast	25.6	32.0	14.1	16.7
	etch*	27.4	35.1	27.2	29.7
EN58J stainless steel	degrease	27.8	31.9	16.7	17.0
	grit blast	27.3	34.0	29.0	25.0
	etch*	27.8	39.9	30.1	33.6
EN3B mild steel	degrease	20.4	24.3	9.8	7.3
	grit blast	23.7	27.4	15.9	18.3

*5 mm etch at 60 °C in 570 g litre^{-1} sulphuric acid, 100 g litre^{-1} oxalic acid.
Source: After refs. 39, 49 and reproduced by kind permission of Elsevier Applied Science Publishers Ltd.

A comparison of the effects of many different procedures is given by Poole and Watts(59), but Fig. 3.11 illustrates the general trend in performance. New processes, developed from the present understanding of adhesion, revolve around maximising micro-mechanical interlocking with the adhesive by 'whisker reinforcement', inhibiting oxide hydration by the use of phosphonate complexes(56), or both. For applications less demanding than those found in the aircraft industry it is possible to achieve reasonable levels of adhesion and durability by gritblasting with alumina followed by the rapid application of a primer or, preferably, a (silane) coupling agent.

Pretreatments for concrete

Surface preparation should include (a) removal of all loose and unsound material until coarse aggregate is exposed (b) cleaning and (c) drying (although there exist some special polymeric systems which are claimed to bond to damp or wet concrete, and cementitious

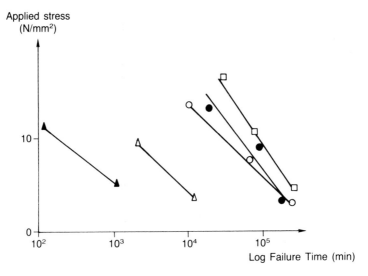

Fig. 3.11. Applied stress versus failure time for nitrile-epoxy/aluminium-alloy joints exposed to 100% r.h. and 52 °C (Ref. 58).

or polymer-modified cementitious systems *must* be used against a dampened surface). Practical problems are likely because of the scale and/or inaccessibility of the works. The importance of cutting back the concrete to a clean sound surface cannot be over-emphasised since adhesion relies partly on mechanical interlock by penetration of the surface pores (see Fig. 3.12), and partly on the physical forces of attraction to clean high energy aggregate surfaces. The comments pertinent to metal oxides apply equally to the hydrated surface oxide layers formed on siliceous materials. There exist several useful publications detailing practical procedures and recommendations(33,60–66), whilst Hewlett(67) gives some detailed consideration to the physical and chemical surface nature of concrete.

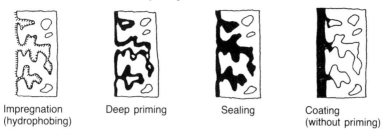

| Impregnation (hydrophobing) | Deep priming | Sealing | Coating (without priming) |

Fig. 3.12. Sealing and coating of concrete surfaces (Ref. 62).

All loose and unsound concrete particles and hardened laitance must be removed, preferably by mechanical means. Subsequent methods for preparation include sand-blasting, bush hammering with blunt tools, needle gun treatment, wire-brushing, water or steam jet cleaning, vacuum blasting and burning off. Chemical etching, with a 10–15% solution of hydrochloric acid, should be used only with previous water saturation of the pore system, to prevent deep penetration of the acid, and the surface should be flushed with water whenever the effervescence ceases. This procedure should be repeated until the surface exhibits a satisfactory texture, but the process is not without its problems. Shields(29) further advises rinsing with water solutions of 1% w/w ammonia, or 10% w/w sodium bicarbonate, to neutralise the acid. Abrasive blasting, too, is not without its drawbacks; aggressive abrasion results in numerous blowholes which may have to be filled prior to coating with adhesive. Cleaning of the substrate surface should follow this removal of the unsound layers. Typical contaminants include grit, dust, oil and greases, and these are best removed by water-blasting, steam-cleaning in conjunction with a suitable detergent, or vigorous scrubbing with a detergent solution or chemical cleaner. Care must be taken not simply to redistribute surface oils and greases.

Finally, the surface must be dried as much as possible, preferably leaving a surface water content of less than 4%. Rough checks might be carried out by holding absorbent paper against the surface, or by taping a polythene sheet to part of the surface and watching for condensation on the underside. Gaul(64) suggests that if this is done under the same ambient conditions and for the same duration as the applied polymer takes to cure, effective bonding may be anticipated if condensation is not observed. The application of penetrating sealers and primers to the substrate may provide the best solution to the creation of a sound and stable surface for bonding(67).

Little information exists on the effects of surface pretreatment on bond strength and durability, and there is a general absence of appropriate test methods to assess such effects on adhesion. The pull-off test(68) is, perhaps, currently the most suitable; the slant-shear test, as described in BS 6319(69), is of limited use in assessing adhesion because the interfaces are not subjected to tensile forces.

Pretreatments for polymer composites

The surfaces of many plastics and rubbers have low surface energies (Table 3.1) such that wetting by an adhesive is inhibited unless special surface pretreatment processes have been employed. However, plastics which contain polar groups such as PVC, nylons and acrylics are bondable with a minimum of surface treatment.

Plastic composites such as glass- or carbon-fibre reinforced materials are often based on polar epoxy or polyester resins, and are therefore compatible with the common adhesives as well as being readily bondable. Surface treatment is required simply to remove contaminants such as oils, dirt, and especially fluorocarbon mould release agents. The two main techniques used to achieve this are:

(a) Solvent wiping, then gritblasting or sanding followed by degreasing.

(b) Provision of a tear-ply at the composite's surface during manufacture, which is stripped off just prior to bonding.

In the former technique the degree of abrasion is known to affect subsequent bond strength and durability, heavy abrasion to expose surface fibres being recommended. Other technologists suggest that 'wet' sanding should be carried out below a 'reactive primer', such as a silane solution, which promotes chemical bonding with the adhesive and carries away the dust from abrasion.

Composite materials, like adhesives, are permeable to water. Moisture can therefore diffuse to the adherend/adhesive interface, which is a problem more in initial joint fabrication when a heat-cured adhesive is employed than when the joint is in service. Thus the moisture content of the composite at the time of joining should be low (say 1%) or else moisture may be drawn to the interface by heating the assembly during curing, so interfering with the adhesion and creating voids in the adhesive layer.

Priming layers and coupling agents

In general, adhesive application to painted surfaces is not to be recommended. However, just as the correct surface treatment is necessary, the application of an adhesive-compatible primer coating may also be desirable. Naturally, the picture of the adhesive–adherend interfacial zone then becomes more complicated. The use of adhesive primers may be more critical in some instances than others but often the advantages to be gained, especially 'pre-wetting' of the substrate surface, far outweigh possible disadvantages such as an extra process, or the primer or a primer interface becoming the weakest link in the joint. The experience of paint and adhesion technologists is that primers greatly reduce the variability of subsequent interfacial bond performance, and that certain products can create a water-stable interface. Their use may also obviate the need for complex surface pretreatment procedures.

As with adhesives, so with primers. The number of candidate products is enormous in order to fulfil any combination of the following requirements:

(1) The immediate coating of a newly prepared surface protects it from damage and contamination. The high surface energies of metals and aggregates are thus converted to ones of much lower surface energy, albeit highly compatible with the adhesive used.

(2) The opportunity to wet the surface more easily and thoroughly than the high viscosity of the adhesive itself would allow, thus facilitating penetration of the surface irregularities and oxide layers.

(3) The ability to block the pores of a porous surface and so prevent capillary suction of adhesive away from the bondline. For concrete such 'bond coats' are often employed, the resin being applied whilst this layer is still tacky.

(4) The provision of an improved mechanical key. In cementitious repair work the 'bond coat', while still tacky, may be dry-dashed with sharp sand.

(5) Corrosion inhibition, which implies treatment of the metallic substrate surface.

(6)   Sacrificial pretreatment, by acting as a hydrophobic 'preferred contaminant' to enable bonding underwater(70).

(7)   The possibility of avoiding the need for complex pretreatments by promoting chemical bonding with coupling agents.

(8)   Prevention of the displacement of the adhesive from the substrate surface by water if chemical bonds can be formed with coupling agents and hydrosols.

In fulfilment of these functions, the two basic approaches which may be discerned are (a) providing a relatively thick barrier or surface coating, and (b) applying a coupling agent to the substrate surface as a monolayer; Fig. 3.13 illustrates the difference. Conventional primers are composed of dilute solutions, 10% solids or less, of the adhesion resin itself in an organic solvent or blend of solvents. Additionally, the primer may incorporate agents to assist wetting, flow control, curing, inhibit corrosion, and toughen the cured primer layer. Light abrasion and solvent degreasing of the primed surface is often advised before application of the adhesive. Hewlett(67) advocates the use of penetrating sealers and primers for concrete surfaces, whilst Hugenschmidt(71) recommends an epoxy–polyurethane coating with a zinc-chromate base for priming steel plates to be used as externally bonded reinforcement in Switzerland. The danger of relatively thick primer layers is that they may become the weak link in a joint because they are themselves mechanically weak. The alternative approach is the use of coupling agents which are used, not specifically to improve bond strength when dry, but to enhance environmental durability. The most common are the siloxanes(72–74) and titanates(75–77), which are applied to metallic or siliceous surfaces in aqueous solutions as a monomer.

Siloxane or silane coupling agents are so termed because they possess a dual reactivity, and are hybrids of silica and of organic materials related to resins. There remains, however, some debate as to the factors affecting the mechanisms by which these materials

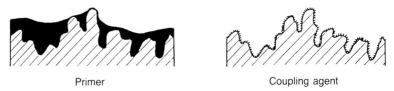

Primer   Coupling agent

Fig. 3.13. Primers and coupling agents.

function(73). Basically, when applied to metal or glass surfaces as a monomer a condensation reaction occurs as it is dried on, leaving a very thin resinous layer which is attached through primary (silicon–oxygen) valency groups to the metal oxide or glass structure. The high energy substrate surface is now hidden and replaced by a surface of relatively low energy. Water displacement of this coating is unlikely since hydrolysis of the silicon–oxygen linkage is a slow process, requiring excess water to be present. In practice, a chemical connection is built into the other end of the silane molecule by inserting organic groups, such as amino-, or epoxy-, which are functionally reactive towards features of the adhesive molecule. It is because of the existence of a truly chemical connection, between oxide surface and siloxane on the one hand and siloxane and adhesive on the other, that these materials are known as coupling agents. Such materials may also be incorporated in the adhesive itself to promote interfacial bonding both with the bulk matrix (resin-filler) and at the adhesive–substrate interface. In GRP technology silanes are used as a matter of course, because water will displace resin from glass; there would otherwise be no fibreglass boats or warships. Comyn(78) cites the case of glass tiles, bonded with epoxy, falling off a London hotel and this problem was subsequently remedied by the application of a silane primer. Figs. 3.14 and 3.15 indicate the superior environmental stability conferred on steel surfaces; when applied to gritblasted steel adherends the interface was found to be completely water-stable.

Gettings and Kinloch(23) used surface-specific techniques to ascertain the bonding mechanisms between silane primer and mild steel, and established that a 1% aqueous solution of Union Carbide's A187 dramatically enhanced joint durability. Hewlett and Pollard(17) examined coupling agents in connection with silicate materials and injection resins, as described earlier, and Allen and Stevens(79) employed infra-red spectroscopy to elucidate the structure of siloxane coupling agents on aluminium. Walker(80), in a series of articles, reported on tests involving a number of silanes, applied to various metallic surfaces and incorporated in a range of paints and coatings. In their application, the four variables to be addressed for any one choice of siloxane are (a) weight of addition (b) composition and pH of the alcohol/water solution (c) time allowed for the mixture to stand before application and (d) time elapsed before application of the adhesive. Due to their hydrophobicity, silane primers are increasingly being used as one of a range of concrete surface 'sealers',

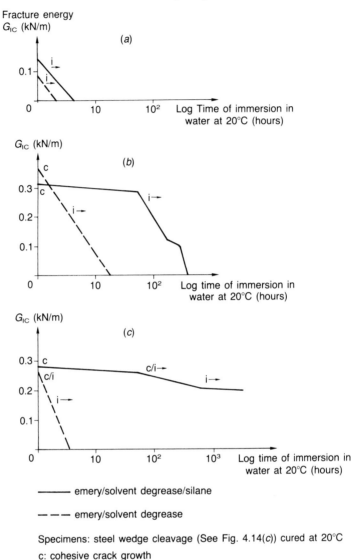

Fig. 3.14. Performance of emery cloth abraded adherends, and the influence of a silane primer, in stressed cleavage tests (Ref. 4). (*a*) Aromatic amine cured epoxy. (*b*) Aliphatic amine cured epoxy. (*c*) Epoxy polysulphide.

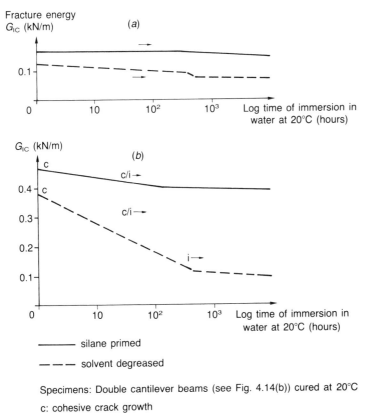

Fig. 3.15. Effect of silane priming of gritblasted adherends prior to bonding in stressed cleavage tests (Ref. 4). (*a*) Aromatic amine cured epoxy. (*b*) Aliphatic amine cured epoxy.

particularly as applied to bridge decks(81,82). The mechanism by which they work is to make the surface cement pores water-repellant, so that salt solutions are not drawn into the body of the concrete by capillary action.

One of the most promising innovations in recent years is the development of hydrophobic Sacrificial Pretreatment Technology (SPT), in conjunction with hydrophobic cold-cure epoxies, to enable underwater bonding(70,83). The energetically-favourable conditions for bonding are established underwater by the application of a water-repellant 'preferred contaminant' to a cleaned or blasted steel

109

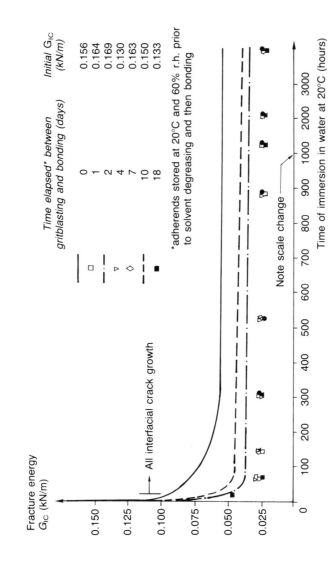

Fig. 3.16. Effect of time lapse between gritblasting and bonding in stressed cleavage specimens constructed with an aromatic amine cold-cured epoxide (Ref. 4).

surface. The adhesive, when applied over this compatible film, either absorbs or displaces it, adheres to the substrate and cures to give a hydrophobic polymer matrix. Burns(84) discusses an equally interesting method of repairing weld failure in oil storage tank floating roof seams, by using an oil and water tolerant cold-curing sealant. A new method of enhancing and maintaining the adhesion between epoxies and steel consists of a process of organic modification of a thin layer (about 10 nm) of tin hydrosol particles, previously deposited onto the substrate from a wetting hydrosol dispersion(85). Corrosion protection and wet peel strengths achieved were superior to all other surface treatments including etching, phosphating and silane application.

The bonding operation

The time elapsed between surface pretreatment and application of the adhesive should generally be kept as short as possible, as implied earlier, because surface energies are lowered by the adsorption of atmospheric moisture and contaminants. Silicone-containing contaminants and mould-release oils should particularly be avoided. Kinloch(7) discusses some theoretical aspects of the bonding operation, and Hewlett and Shaw(60) address issues pertinent to segmental bridge construction. For general construction applications the considerations would involve (a) surface cleanliness and dryness (b) adhesive useable life (c) ambient temperature (d) adherend temperature (if different from (c)) and (e) minimising air entrapment. Note that the adherend temperature should ideally not be lower than that of the surrounding air prior to application of the adhesive (and/or primers), in order to prevent moisture condensing at the interface; local heating may be advisable, which would also assist surface wetting. Brockmann(41) reports little effect on the strength of mild steel–epoxy lap joints due to storing the shot-blasted adherends at room temperature and 60% r.h. for several days prior to adhesive application. This should not occasion surprise since a shear test is insufficiently sensitive and is inappropriate for gauging effects on adhesion. Gledhill *et al.*(19), on the other hand, found that the effect of bonding in a high humidity environment was reflected in reduced tensile butt-joint strengths. Fig. 3.16 also demonstrates a clear time-lapse effect on fracture energy.

3.4 Methods to study pretreatments

The wide range of techniques for studying pretreatment effects may be classified as (a) mechanical (destructive) and (b) surface analytical (non-destructive). Naturally, different substrate materials warrant different approaches.

Mechanical test procedures

Experimental assessments of the effects of surface treatments using mechanical tests are of limited value unless environmental exposure is included. Appropriate comparative tests subject the interface to tensile, peel or cleavage forces but unfortunately most of these, including the pull-off test for adhesion(86,87), employ joints in which adhesive and adherend strains are far from uniform so that the results reflect an interaction of stress concentrations. Further, few of the joint configurations allow environmental access within a reasonable time-scale of exposure. However, the self-stressed fracture mechanical cleavage specimens described in Chapter 4 (4.14(band c)), and employed in most of the experimental work referred to in this chapter, do afford a quantitative means of assessing pretreatments for metals and the effects of environmental exposure in terms of measurable adhesion (2,4,5,88). The adherends are forced apart and fracture is calculated from a knowledge of the adherend displacement, and the crack length measured by travelling microscope. The location of the crack-tip is very important, because a crack will transfer rapidly from the bulk adhesive to an unstable interface. Initial crack propagation is generally rapid initially, and then stabilises to indicate a threshold fracture energy for the particular environment of exposure.

Surface analysis

There is a wide variety of techniques available(10,89):

(a)	Contact angle	to compare surface energies
(b)	Optical and electron microscopy	to examine surface topography
(c)	Surface analytical methods	to study surface chemistry, particularly within the first few angstroms of surface.

Contact angle measurements, as discussed earlier, are limited in their applicability to ideal surfaces or to surfaces wetted by liquids of homogeneous composition resting in equilibrium. The nature of real surfaces, real adhesives and the pressure of bonding cannot be allowed for as encountered in real bonding operations. Nevertheless, some comparative information on surface energies and the wetting of some real surfaces may be anticipated.

Surface topography and morphology may be studied with optical or electron (scanning or transmission) microscopy, and these techniques are very useful for monitoring the physical changes due to pretreatments (e.g. Fig. 3.7). Indeed the significance of micromechanical adherend–adhesive interlocking has only recently been appreciated as an important mechanism of adhesion for bonds to certain adherends. Transmission Electron Microscopy (TEM) gives even greater magnification than Scanning Electron Microscopy (SEM), but the former technique involves the preparation of replicas.

From many potential techniques to study surface chemistry, only a few are of particular importance. For metals Auger Electron Spectroscopy (AES), Electron Spectroscopy for Chemical Analysis (ESCA) also known as X-ray Photoelectron Spectroscopy (XPS), and Secondary Ion Mass Spectroscopy (SIMS) give useful information about the chemistry within the first nanometre of surface. ESCA (or XPS) is also used in the study of polymer surfaces, as are infrared techniques such as Multiple Internal Reflection IR and Fourier Transform IR. Inelastic Electron Tunnelling Spectroscopy (IETS) is a way of obtaining the vibrational spectra of molecules adsorbed on a metal oxide, and the technique has been used to examine adhesives and coupling agents. Good evidence of primary chemical bonds between certain silane coupling agents and both substrate and epoxide groups has been found(90,91).

3.5 Summary and concluding remarks

Optimisation of surface pretreatment is the key to maximising joint durability. The adhesive influences the surface oxide layer and the surface oxide layer influences the boundary layer polymer matrix; the whole must therefore be viewed as a unique system for every adherend–adhesive combination. The interplay of chemical bonding and microstructural and macrostructural behaviour greatly complicates the study of adhesion. The chemistry and the mechanics of adhesion are not independent, so that the intrinsic forces of adhesion cannot be equated with measurable adhesion.

The role of surface energies and the mechanisms of adhesion are essential to an understanding of favourable conditions for wetting, and of appropriate surface treatments. Adhesives can be encouraged to wet most adherends if their surfaces are suitably prepared to make them 'receptive'. Unfortunately, high energy surfaces are readily wetted by atmospheric moisture and airborne contaminants, both prior to bonding and also after the adhesive has cured if these agencies can migrate to the interface. Thus lower energy hydrophobic substrate surfaces are preferred, such as primed surfaces.

In the practice of adhesive bonding for applications in construction, surface pretreatment is likely to be the most difficult process to control. The choice of treatments must be tempered by the scale of operations, the nature of the adherends, the required durability, the adhesive to be used, and the cost. The performance of joints constructed with cold-cure epoxies is likely to be critically dependent upon surface preparation, as exemplified by the experience of the Scottish Irvine Development Corporation. In 1978 they elected to use vertical externally-bonded steel plate reinforcement to strengthen the abutment walls of three pedestrian underpasses. A year later, the plates were reported to be falling off, accompanied by extensive interfacial corrosion; the steel surfaces had been abraded by hand, and the concrete surfaces chemically etched.

Some adhesives, notably the acrylics, the heat-cured epoxies and the plastisols are able to withstand a certain amount of substrate contamination. At room temperature the solubility of oil in all epoxy resins is low, but at higher temperatures appropriate formulation can make these materials into reasonable solvents. Plastisol adhesives, which have been used widely in automobile construction on unprepared steel surfaces, differ because the plasticising oils they contain become very powerful solvents as the curing temperature (say 180 °C) is approached. During curing these oils and the contamination they pick up are incorporated in the hardening adhesive mass. However, the use of apparently tolerant adhesives is not an alternative to good surface preparation, because the pretreatment undoubtedly plays a significant role in subsequent joint durability.

Generally, the adherend surfaces should be clean, dry, and free from oils, grease and all loose and unsound material. Adhesives join materials by attaching to their surface layers, so that it is important to think in terms of bonding, at molecular scale, to oxide layers (which are generally inherently unstable!). Bonds to metal alloys and thermoplastics are problematical but have, in fact, been the subject of intensive research because they are commonplace adherends. Currently, little is known about optimising concrete surface treatments. Indeed for many practical adhesive–substrate interfaces there remain unresolved debates concerning the detailed mechanisms of adhesion and of environmental failure. Despite the extra processes involved, hydration-resistant surface treatments and coatings, coupling agents, or primers would seem to be very worthwhile both in reducing variability in joint performance and in securing environmentally stable interfaces. In recent years, the development of new hydrophobic adhesive formulations and sacrificial pretreatment technology have become of considerable interest for bonding under 'difficult' conditions.

Experimental assessments of the effects of surface pretreatment are of limited value using mechanical tests unless environmental exposure is included. It is very sound policy to collect and examine information on joints loaded and exposed to natural weathering conditions rather than depend solely on laboratory experiments. It is clear that water is the substance which causes most problems in attaining environmental stability of bonded joints; interfacial failure generally indicates that a better surface pretreatment would impart improved joint performance.

CHAPTER FOUR

Adhesive joints

4.1 Introduction

The truly structural adhesive joint is relatively new. The evolution of the various design approaches follows the empirical development of appropriate joint configurations(1–7) – themselves following on from the long historical development of load-bearing joints in, and between, engineering materials. It must however be emphasized that structural bonded joints existing in engineering disciplines other than those involving civil engineering tend to be formed with *thin* bondlines, often with relatively high modulus adhesives, whereas the general concern in the construction industry is with *thick* bondlines – often with lower modulus materials. This is an important difference, since the nature of the resultant bondline stress distributions of loaded joints may be significantly different.

The training normally given to an engineer in the various means of joining materials leaves him at a disadvantage when it comes to using adhesives, with essential choices between the many types available and with the design approach appropriate to structures assembled with these. Naturally the basis for design must stem from the intended function and service environment of the joint, and from a consideration of the loads and stresses which are likely to be encountered in service. As with any fastening method, it follows that the design must be dependent upon the nature of the materials to be joined as well as on the method of joining. It is, for instance, not sufficient simply to substitute adhesive bonding for welding, bolting or riveting.

The properties of the composite made when two adherends are united by adhesive are a function of the bonding, the materials involved and their interaction by stress patterns. Potential problems implied by the latter stem from the inherent mismatch between adhesives and the materials commonly employed in construction (Table 4.1). For instance, concrete adherends would benefit from being united with flexible and relatively low modulus products in

116

Table 4.1. *Comparison of typical properties*

Property (at 20°C)	Cold-curing epoxy adhesive	Concrete	Mild steel
Relative density	1.3	2.2	7.8
Young's modulus (GN/m^{-2})	4	30	210
Shear modulus (GN/m^{-2})	1.4	10	80
Poisson's ratio	0.37	0.18	0.29
Tensile Strength (MN/m^{-2})	25	4	400
Shear strength (MN/m^{-2})	30	5	550
Compressive strength (MN/m^{-2})	75	40	400
Tensile elongation at break (%)	0.5–5	0.15	30
Approximate work of fracture (J/m^{-2})	100	20	10^5–10^6
Linear coefficient of thermal expansion, per °C	35	10	11
Water absorption 7 days at 25°C (% w/w)	1	5	
Glass transition temperature (°C)	45		

order to reduce interfacial stress concentrations which might initiate fracture of the concrete. On the other hand, sustained loading may lead to excessive deformation and creep unless the adhesive used is relatively unmodified and therefore highly cross-linked, while also possessing a high glass transition temperature. However, thin sheet metal, metal alloys and composite materials demand either a

117

toughened product or a flexibilised adhesive with a large strain to failure in order to accommodate gross adherend strain under load.

It is apparent that in order to realise optimum performance from a load-bearing joint, balance between the requirements of the different materials to be joined can only be achieved by rational design which requires, *inter alia*, quantitative data on the properties of structural adhesives. Currently there is an acute lack, both of adhesives performance data and of appropriate test procedures for determining relevant structural properties. Indeed, because so many factors affect joint strengths, extreme care must be taken when interpreting published performance data. For instance, details such as joint configuration, testing conditions or surface treatments may be insufficiently described to make comparisons of results collected from different sources useful. Appropriate test procedures, with a view to the long-term performance as well as to the short, are therefore discussed in this chapter.

It has been emphasised already that a successful adhesive bonded joint depends upon several factors:

(1) appropriate design of the joint
(2) selection of a suitable adhesive
(3) adequate preparation of the adherend surface
(4) controlled fabrication of the joint
(5) protection of the joint itself from unacceptably hostile conditions in service
(6) post-bonding quality assurance.

The significance of some of these factors, particularly surface pretreatment, tends to become more apparent with regard to durability and to long-term performance.

Kinloch(4) observed that the selection of appropriate failure criteria for the prediction of joint strength by conventional analysis is fraught with difficulty. The problem is in understanding the mechanisms of failure of bonded joints, and in assigning the relevant adhesive mechanical properties. Current practice is to use the maximum shear-strain or maximum shear-strain energy as the appropriate failure criterion. However, the failure of 'practical' joints occurs by modes including, or other than, shear failure of the adhesive. This difficulty has led to the application of fracture mechanics to joint failure.

118

4.2 Factors affecting joint strength

The strength of a bonded joint will be determined by the strength of its weakest component, which is generally designed to be the adhesive. It follows that the requirements for satisfactory joint performance are (a) good contact between the adhesive and substrate (b) absence of weak layers in the joint (c) that the adhesive should possess appropriate mechanical properties. These basic requirements are implicit in the essential elements of bonding, and are related to many factors which affect the performance of bonded assemblies as summarised in Table 4.2.

Table 4.2. *Factors affecting joint 'strength'*

(1) Joint design
geometrical configuration
bondline thickness

(2) Adherends
mechanical properties
susceptibility to deterioration
linear coefficient of thermal
    expansion
permeability

(3) Adherend surface
surface chemistry
surface topography
surface cleanliness

(4) Nature of primer (if applicable)
viscosity
chemical composition
mechanical properties

(5) Nature of coupling agent
    (if applicable)
chemical functionality
dilution factor in solution

(6) Nature of adhesive
rheology – viscosity
chemical composition
reactivity – pot-life
mechanical properties
linear coefficient of thermal
    expansion
resistance to biodeterioration
permeability

(7) Bonding conditions
temperature of substrate
ambient temperature
humidity
air-borne contamination
open time
cure time
pressure

(8) Internal stress
cure shrinkage
temperature
environmental conditions
nature of adherends
nature of adhesive

(9) Service/environmental conditions
stress
moisture
temperature

(10) Testing conditions
strain rate
cyclic frequency
temperature

Given that the adhesive itself should determine the strength of a bonded joint, the stress required to rupture a joint is, nevertheless, not a well-defined materials constant. When two materials are bonded, the resultant composite has at least five elements, namely the adhesive itself, two adhesive/adherend interfaces, and two adherends. If a primer is applied to both substrate surfaces, the number of elements increases to at least nine. These elements involving a metallic adherend are depicted schematically in Fig. 3.2. Note that the adhesive (or primer in this case) is in contact with the metal surface oxide layer, and not with the metal itself.

Kinloch(8) suggests that the measured bonded joint strength almost always reflects the value of two parameters:

(1)   the intrinsic adhesion
(2)   the energy dissipated visco-elastically and plastically in the highly strained volume around the tip of the propagating crack and in the bulk of the joint.

The latter term generally dominates the measured joint strength, and also gives rise to the test rate and temperature dependence of joint strengths. Brittle fracture may be initiated at the interface with unmodified (rigid) adhesives if the energy of fracture cannot be dissipated within the adhesive layer.

Real joints do not of course consist of simple, separate, elastic materials with a clear mathematical geometry. Metal adherend surfaces are micro-rough, possessing oxide layers, while concrete surfaces are macro-rough comprising aggregates and cement paste, and both surfaces readily adsorb air-borne contamination. The thickness and modulus of primer layers, if employed, is often unknown, and the thickness and properties of the adhesive layer are difficult to regulate and to determine.

In order to develop interfacial strength the adhesive must be involved in wetting, adsorption and inter-diffusion reactions with the adherend. Its chemical composition will influence the extent of interaction and its ability to displace and absorb surface contamination. The adherend surface topography can affect joint strength in several ways. Irregular surfaces have a greater potential bonding area than smooth surfaces, and mechanical keying may play an important role. However, wetting may be far from complete with viscous adhesives, particularly if the irregularities are deep and narrow; resultant voids could act as stress raisers. The viscosity of the adhesive may be lowered by increasing the temperature, and wetting will also be enhanced. Primers may also be used to overcome the wetting problem. Rheological aspects of adhesion, together with the simultaneous influence of adherend surface chemistry on wetting and adhesion, are considered in Chapter 3. Adherend surface pretreatment is also described in the previous chapter, in particular because of its profound influence on bond durability.

The reader will appreciate that a large number of factors can affect joint strengths, and hence the caveat advanced in the introduction that extreme care must be exercised in the interpretation of bonded joint performance.

4.3 Joint design

General considerations

The adhesive, which can be likened to plastic material, represents a low modulus interlayer and is likely to be the weakest link in a structural joint. Exceptions to this might be when one or both substrates are concrete subjected to forces other than compression, or if thin sheet metal adherends are involved. Design considerations generally involve the geometry of the bond, the selection of an adhesive and the necessary bonding process, a knowledge of the properties of the adhesive and of the adherends, and finally an analysis of the stresses that the bonded assembly is likely to encounter in service. Petronio(9) states that for maximum success joint design should follow several general principles, namely to:

(1)   stress the adhesive in the direction of maximum strength (i.e. in compresion or in shear)
(2)   provide for the maximum bond area
(3)   make the adhesive layer as uniform as possible
(4)   maintain a thin and continuous bondline
(5)   avoid stress concentrations.

Assuming that the essential factors necessary to create a reliable joint as outlined in the introduction have generally been complied with, joint strength is related essentially to the mechanical properties of the adhesive with which it is constructed. Since most of the adherends united by structural adhesives are loaded in tension, the subsequent loads and hence stresses on the adhesive are then a function of the geometry of the joint. Unfortunately even in the simplest of joint configurations the stress state is complex, and Adams and Wake(5) give very detailed consideration to the nature and magnitude of stresses in joints.

Most of the adhesives used for structural purposes are relatively strong in shear, and weak in peel or cleavage. Bonds are therefore usually designed to place the adhesive in shear and to minimise these other stresses. Excluding compression, the four important types of stress to consider are shear, tension, cleavage and peel, and the resulting joint elastic stress distributions are shown schematically in Fig. 4.1. (Note that peel is the enemy of the joint designer and implies principally transverse tearing loads.) Kinloch(4) presents a useful chronological review of the approaches made in analysing the stresses in various kinds of bonded joint.

The lapped joint depicted in Fig. 4.1(*a*) is one of the most commonly occurring joints in practice and is therefore the configuration most often used for testing adhesives(10, 11). As shown, the problem is that the loads are not linear, and a bending moment must exist so causing the joint to rotate. In consequence, the adhesive layer will be subjected both to shear and to tearing stresses at the ends of the joint. The adherends, too, are subjected to shearing, stretching and bending. As the load is increased it is probable that both adhesive and adherend will become plastic in the highly-stressed regions. The double lap joint, whilst eliminating gross joint rotation, is really no more than a back-to-back arrangement of two single-lap joints and significant tearing stresses still exist.

Lapped joints, howsoever fastened, support most of the applied load as stress concentrations at the two ends of the joint, and it is for this reason that the strength of such joints depends largely upon their width and very little upon the overlap length. So far as relatively rigid lapped joints are concerned Gordon(12) observed that it makes little difference whether the joint is glued, nailed, screwed, welded, bolted or riveted. However, as the imposed load on bonded lap joints is raised the adhesive layer may fracture before it yields. This problem may be overcome to a certain extent by using flexibilised adhesives which yield considerably before they fracture. Alternatively, toughened adhesives may be employed to inhibit crack propagation through the bondline, and so prevent sudden fracture. These approaches are not without their drawbacks because flexibilised and toughened adhesives may creep readily under sustained load. Further complexities of the problem such as

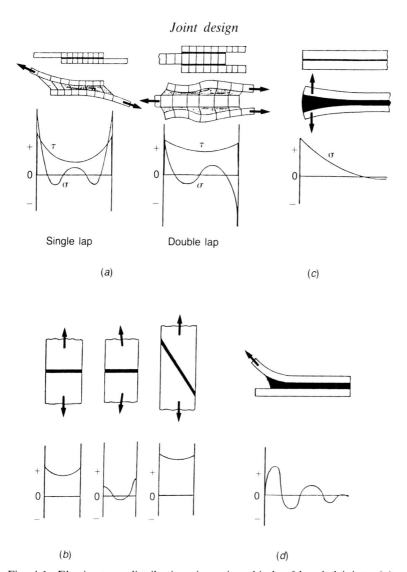

Fig. 4.1. Elastic stress distributions in various kinds of bonded joints. (*a*) Lap shear. (*b*) Butt-tensile and scarf. (*c*) Cleavage. (*d*) Peel.

the adherend surface rugosity, cure shrinkage, the flaw-sensitive nature of cured adhesives, and the adhesive in-bondline mechanical performance are also apparent. On the positive side, real bonded joints are inevitably made with spew fillets of adhesive at their extremities and, as studied extensively by Adams and co-workers

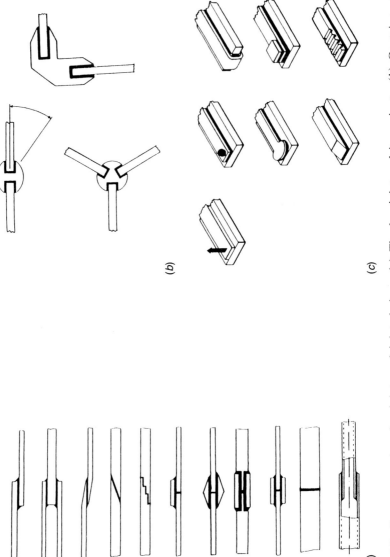

Fig. 4.2. Common engineering joints and joint designs. (a) The lap joint and its variants. (b) Containment joints for plates, extrusions and pultrusions. (c) Ways of minimising peel in laps, doublers and stiffeners.

(a)

(b)

(c)

at Bristol University during the 1970s and 80s, this generally has the effect of reducing stress concentrations at the end of the overlap. The adherends may also be made less stiff at the ends of the joint by tapering, scarfing and so on, to smooth the transfer of stress and strain across the joint. Many such examples of good practice in joint design for 'production' applications are depicted in several texts(2, 5, 6, 7), and some of these are shown in Fig. 4.2. The scarf joint, though ideal in minimising stress concentrations, is seldom practicable.

The design of a structural adhesive joint intended to resist significant loads is clearly quite onerous. For instance, it is not permissible to assume uniform stress distributions because at ultimate conditions fracture is often experienced. With traditional methods of fastening the same stress concentrations will exist but the materials can normally be relied upon to yield and there is little need to be so concerned about transverse normal tensile stresses. In order to make reliable strength predictions, bonded joint design necessitates either some form of mathematical analysis, or else extrapolation from experimental data. Two such design philosophies are based on stress analysis and on fracture mechanics. It is assumed implicitly that other modes of joint failure such as impact, excessive deformation, creep and gross corrosion are considered in parallel with these design techniques.

Stress analysis and design philosophies

Closed form solutions. Structural adhesive joints are generally designed to be loaded in shear so that treatments of joint analyses are confined essentially to the transfer of load by shear, with some consideration of the transverse normal stresses induced by eccentricities in the load path. In the simplest case the adhesive and the adherends are assumed to behave elastically. The most refined analyses attempt to model the situation when the adhesive yields so that the adhesive and, eventually, the adherends behave plastically as the imposed load is raised. Closed-form analyses are difficult to apply to other than simple geometrical configurations, while a major difficulty with the elasto-plastic model is how to characterise the adhesive.

The empirical approach to the design of simple overlap joints was to construct a correlation diagram (Fig. 4.3) between failure load

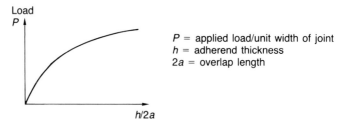

Fig. 4.3. Empirical correlation diagram.

and the joint geometrical ratio $h/2a$, for a particular set of test conditions. Alternatively, if the apparent shear stress at failure of a series of lap joints is plotted against the 'joint factor' $\sqrt{h/2a}$, as recommended by de Bruyne(13), an approximately straight line is obtained over a wide range of adherend thicknesses (h) and overlap length ($2a$). This factor works well with adhesives that give ductile failures, but has limited applicability to rigid adhesives. Inspection of the stress concentration coefficient, Δ, as derived by Mylonas and de Bruyne(14), indicates the presence of the 'joint factor':

$$\Delta = \frac{4G_a a^2}{Eht} \tag{4.1}$$

where

G_a = shear modulus of adhesive

E = Young's modulus of adherend

t = thickness of adhesive

This expression correctly shows that the overlap length should depend upon the modulus of the adhesive.

The mathematical treatment of joint analysis is to set up a series of differential equations to describe the state of stress and strain in a joint. By using stress functions or other methods, closed-form algebraic solutions may be obtained. In the simplest elastic case it should be possible to devise a solution for given boundary conditions. As non-linearities arise, such as joint rotation and material plasticity, various assumptions need to be made to give solutions. However, once obtained, these solutions may be used to great advantage in a parametric study, provided the limits of the simplifications are borne in mind. The classical early work of Volkersen(15) and of

126

Goland and Reissner(16) was limited because the peel and shear stresses were assumed constant across the adhesive thickness, the shear was assumed a maximum at the overlap end (and not zero as it must be at a free surface), and the shear deformation of the adherends was neglected. Later, several analysts including Renton and Vinson(17) and Allman(18) produced solutions where the adherends were modelled to account for bending, shear and normal stresses. Adhesive shear stress was set to zero at the overlap ends. Allman additionally allowed for a linear variation of the peel stress across the adhesive thickness, although his adhesive shear stress was constant.

Later, extensive studies were carried out by Hart-Smith(19) during the Primary Adhesively Bonded Structure Technology (PABST) programme, which ran from 1976 to 1981, to account for adhesive elasto-plastic behaviour. He developed many computer programs for analysing various joint configurations, requiring the input of an idealised adhesive stress–strain curve (Fig. 4.4(*a*)). Similar programs are available from the Engineering Sciences Data Unit (ESDU) in London for elastic(20) and inelastic(21) calculations, developed from the work of Volkersen and Goland and Reissner. In Hart-Smith's work, yield stress was effectively equated to the failure stress so that failure occurs when the adhesive reaches its limiting shear strain. By increasing the overlap length, the lightly stressed region in the middle of the joint is enlarged. Current aerospace design philosophy views this as essential to offset the effects of creep at the ends of joints if low cycle creep/fatigue loading is encountered

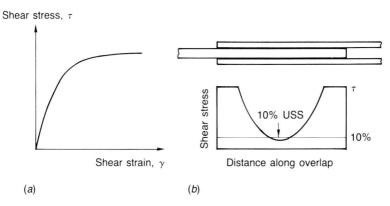

Fig. 4.4. Design of bonded lap joint (Ref. 19). (*a*) Idealised adhesive shear stress/strain curve. (*b*) Design of bonded double-lap joints.

127

(e.g. the pressurisation/de-pressurisation cycles encounted by aircraft fuselages). Fig. 4.4(*b*) summarises Hart-Smith's design criteria for double-lap joints. The overlap is designed for the worst service condition, which is usually when the adhesive has been softened by moisture and elevated temperature. The plastic zones are calculated to be long enough to carry the ultimate load, the central elastic region large enough to prevent creep, and the minimum operating stress must not exceed 10% of the adhesive shear strength. In the PABST programme a ratio of overlap length to adherend thickness of 80 : 1 was chosen for single-lap joints. Armstrong(22) reports that Boeing used 50 : 1, but that Douglas prefer a ratio of 80 : 1. It should be borne in mind that these design philosophies were developed specifically for aircraft construction involving the bonding of thin metal alloys. Whilst representing a significant advance in the design of structural adhesive lap joints, they may not be generally applicable to other engineering applications.

Pera International, in collaboration with France's Centre Technique des Industries Mécaniques (CETIM), developed an approach to the prediction of bonded joint strength during a four-year (1986–90) project called Adhesive Bonding Technology for Engineering Applications (ADENG) funded through the European Community's BRITE/EURAM programme. The approach, whilst based upon the use of software employing closed-form analyses, represents a very useful design tool of general applicability to bonded joints encountered commonly in production engineering. Other useful software design techniques under the names of BISEPS-TUG (Bonded Inelastic Strength Prediction Suite – Tubular Geometry) and – LOCO (Lap Joint Combined Loading) have been developed by the Materials Development and Engineering Divisions of Harwell Laboratory, also largely within European Community-funded programmes during the 1980s. CADEPT, an expert system for the design of adhesively bonded coaxial joints, has also been developed more recently by the Harwell Laboratory. Some rudimentary analytical capability also forms part of Permabond Adhesives' Locator (PAL) software.

The methods developed over the last decade or so to predict failure employ in-bondline non-linear adhesive characterisation. This in turn requires some fairly sophisticated experimental techniques, carefully conducted for a range of test conditions and environments. Hart-Smith concluded that a precise representation of the adhesive stress–strain characteristic is not important. He maintains that the adhesive failure criterion, in shear, is defined uniquely by the adhesive shear strain energy per unit bond area, regardless of the individual characteristics which contribute to that strain energy.

Finite element analysis. The finite element method is now a well-established technique for modelling mathematically stress analysis problems. Its great advantage lies in the fact that the stresses in a body of complex geometrical shape under load can be determined. The method is therefore well suited for analysing and optimising adhesive joint geometries, particularly when formed with adhesive spew fillets at their extremities. The method also avoids the approximations 'of the closed-form theories, thus enabling more accurate answers to be found. However, because of stress gradients both across and along the adhesive layer, it is necessary to use quite a large number of elements to give sufficient resolution even in standard model lap joints.

Adams and co-workers have been amongst the main proponents of this method in its application to a realistic analysis of bonded lap joints(5). Harris and Adams(23) presented a non-linear analysis that was able to take account of large displacement joint rotations and elasto-plastic properties of both adhesives and adherends. It was used to predict the mode of failure and ultimate load of single lap joints constructed with aluminium adherends and epoxy adhesives. The results were compared with those obtained from experiment and closed-form analyses. They found that a failure criterion based on the triaxial tensile properties of the adhesive (which they then approximated to the uniaxial tensile properties) could be assigned to these particular joints. For the untoughened adhesives a maximum stress criterion was appropriate, while for the toughened adhesives a maximum strain criterion could be employed. Again, fairly sophisticated adhesive characterisation was required, both for the analysis and for the assignment of the failure criteria. Further development of Harwell's BISEPS software has added some finite element capability to their suite of closed-form programs.

For investigations into real joints the finite element technique is very powerful. The trouble is that each solution applies only to a given set of parameters so that many computer runs are required for a parametric analysis. Even with mesh generation programs, the cost in mainframe computing time can be rather high if it is necessary to model the structure with a large number of elements; however, the advent of powerful PCs may soon render rapid and economic analysis practicable.

The fracture mechanics approach. Traditional structural design in conventional materials compares average strain, and therefore average stress, distributions acting on some ultimate criteria. For designs involving brittle flaw-sensitive materials such procedures are less satisfactory. Adhesive joints usually fail by the initiation and propagation of flaws and, since the basic tenet of continuum fracture mechanics is that the strength of most real solids is governed by the presence of flaws (Griffith(24)), these theories have proved to be extremely useful in their application to adhesive joints. Comprehensive review material related to adhesive joints is given by Kinloch(4), Anderson *et al.*(25), Bascom *et al.*(26, 27) Kinloch and Shaw(28), and Ripling *et al.*(29). Basically, two inter-relatable conditions for fracture have been proposed.

First, the energy criterion for fracture, which is simply an extension of Griffith's hypothesis. He postulated that fracture occurs when sufficient strain energy is released to satisfy the requirements of the new surface area created at the instant of crack propagation. In fact, crack initiation involves a number of energy consuming processes within the immediate vicinity of the crack tip so that the Griffith equation requires some modification, as proposed by Irwin(30). This approach therefore provides a measure of the energy required to extend a crack over unit area and is termed the fracture energy or strain-energy release rate, G_c.

Second, Irwin found that the stress field surrounding a crack could be defined uniquely by a stress-field parameter termed the stress-intensity factor, K. He postulated that fracture occurs when the value of K exceeds some critical value, K_c, often referred to as the material fracture toughness. Thus K relates the magnitude of the stress-intensity local to the crack in terms of the applied loadings and the geometry of the structure in which the crack is located. A crack in a solid may be stressed in three different modes as depicted in Fig. 2.18. Mode I opening, and hence the Mode I value for the stress intensity factor K_I, is the most critical situation in bonded joints.

Hence the condition

$$K_I \geqslant K_{IC} \tag{4.2}$$

represents a fracture criterion. It may be related to the geometry of any particular structure via the equation

$$K_{IC} = Q\,\sigma_c\,(\pi\,a_f)^{\frac{1}{2}} \qquad\qquad (4.3)$$

where Q is the configuration correction factor as a function of geometry and loading, σ_c is the stress at crack initiation, and a_f is the flaw size.

Pure Mode I fracture energy is known to be considerably less than either of the other modes, but many practical situations involve combined-mode loading about which little is known(31). Bascom and Hunston(27) demonstrated that the use of Mode I fracture energies in design was not conservative because of mixed-mode effects, especially where elastomer-modified adhesives are to be employed. Various test procedures have been devised to obtain values for G_{IC} and K_{IC} for an adhesive, from both bulk and joint specimens. Like other material properties these are not unique values, but rather are dependent upon the rate and temperature of testing, environmental conditions, and the geometry of the specimen. Constraining the adhesive as a thin bondline between high modulus substrates may also complicate the problem because of its influence on the development of the plastic zone (Fig. 4.5). Further, Bascom *et al.*(26) suggests that the size of this zone is very much larger than that associated with untoughened adhesives.

The relationships between G and K for a crack in a homogeneous material are given by:

$$K_I^2 = EG \qquad \text{for plane-stress} \qquad\qquad (4.4)$$

$$K_I^2 = \frac{EG}{(1 - v^2)} \qquad \text{for plane-strain} \qquad\qquad (4.5)$$

For a crack in an adhesive layer these relationships are still generally valid so that K (joint) and G (joint) may be correlated through E (adhesive), albeit approximately in the case of very thin adhesive layers. For a crack near an interface though, the interpretation of K is more complex, probably involving values for K_I and K_{II}.

A major advantage of the fracture mechanics approach to joint failure is that it is applicable equally to cohesive and adhesive failure

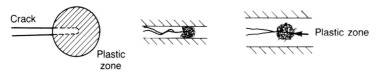

Fig. 4.5. Change in crack tip deformation zone with increasing bondline thickness (Ref. 27).

131

modes. The significance of this advantage in elucidating the mechanics and mechanisms involved in joint failure will be apparent. Indeed, the success of this approach is demonstrated by the widespread adoption of test methods based upon fracture mechanics principles in the aerospace industry.

The significance of the fracture mechanics approach to joint design is that G_c or K_c may be related to a critical crack length in the structure. Cracks which are shorter are safe, and cracks which are longer are self-propagating and potentially catastrophic. Thus structures may be designed to accommodate cracks of a pre-determined length without breaking. This length naturally has to be related to the size of the structure and also to the probable service and inspection conditions. The fail-safe design concept, according to Parker(31), assumes that in spite of the failure (or incipient failure) of an individual component, the complete structure is safe from overall catastrophic failure. Practically, such designs may be achieved by (a) multiple load paths encouraged by stringers (b) crack arrestors to inhibit propagation (c) routine inspection. In bonded joints the use of toughened adhesives to inhibit crack propagation is an attractive solution, so long as fracture is energetically favoured within the adhesive layer, and provided that other performance aspects are not compromised unduly by the toughening.

4.4   Test procedures

General remarks

It is evident that a large number of parameters are involved in the fabrication and testing of bulk adhesive specimens and adhesive joints; these must be controlled if meaningful experimental data are to be obtained. Joint tests evaluate not only the mechanical properties of the adhesive, but also the degree of adhesion and the effectiveness of surface treatments. The standard test procedures listed by ASTM, BSI, DIN and other official bodies are essentially for testing adhesives and surface treatments rather than joints (e.g. Table 4.3). Unfortunately, most of these tests consist of joints in which the adhesive stresses are far from uniform. The designer and the researcher therefore have to select appropriate tests, and to know what the results mean in terms of their own particular investigations and applications.

Adhesive tests may be used for various reasons, including(25):

(1)   comparing the mechanical properties of a group of adhesives
(2)   as a quality check for a 'batch' of adhesive
(3)   checking the effectiveness of surface and/or other preparations
(4)   as a means of determining parameters that can be used in predicting performance in actual applications.

(1)–(3) may be classed as qualitative tests and a number of procedures may suffice. Quantitative results as required for reason(4) are much more demanding, requiring a detailed knowledge of the stress distribution within the test joint and its relationship to that in a real joint.

Structural analysis requirements

Quantitative data requirements for stuctural analysis depend on the design approach and, therefore, the type of analysis to be performed. However, all the theoretical methods for predicting joint strength require:

(1)   adherend tensile modulus (E) and Poisson's ratio (v)
(2)   adhesive shear (G_a) and tensile (E_a) moduli, and Poisson's ratio (v_a)

Additionally, analyses which allow for adhesive non-linear behaviour will need data on ductility, e.g.:

(3)   yield stress (strain) and ultimate stress (strain), in shear or tension, or both

Some analyses require adhesive physical properties such as:

(4)   coefficients of thermal and hygroscopic expansion.

A number of analysts (e.g. 5, 19) maintain that non-linear analysis is the key to being able to predict failure of bonded joints. It is also apparent that the strength of bonded joints, however loaded, is determined largely by the ultimate stress or the ultimate strain capability of the adhesive in tension.

Table 4.3. *Some standard test methods for adhesive-bonded joints*

Test	Standard		Remarks
Definitions	ASTM D907-82(85)		Standard definitions of terms relating to adhesives
Axially-loaded butt joints	BS 5350: Part C3: 1978 ASTM D897-78(83)		
	ASTM D2094-69 (80) and D2095-72(83)	)	Specifically for bar- and rod-shaped specimens
	BS 5350: Part C6: 1981	)	Bond strength in direct tension in sandwich panels
Lap joints loaded in tension	BS 5350: Part C5: 1976 ASTM D1002-72(83)		Single- or double-lap joint test Basic metal-to-metal single lap joint test
	ASTM D3528-76 (81)		Double-lap joint test
	ASTM D3163-73(84) and D3164-73(84)	)	Specifically for polymeric substrates
	ASTM D2295-72 (83)		Single-lap joint test for metal-to-metal joints at elevated temperatures
	ASTM D2557-72(83)		As above but at low temperatures
	ASTM D905(86), D906(82) D2339-82 and D3535-79(84)	)	Specifically for wooden joints
	ASTM D3983-81(86)	)	Thick substrates used; shear

Test	Standard	Description
	DIN E54451-77	modulus and strength of adhesive determined
Peel joints	BS 5350: Part C9: 1978 and ASTM D3167-76(81)	Floating-roller test
	BS 5350: Part C10: 1979 and BS 5350: Part C14: 1979	90° peel test
	BS 5350: Part C11: 1979 and ASTM D903-49(83)	180° peel test
	BS 5350: Part C12: 1979 and ASTM D 1976-72(83)	'T' peel test for flexible-to-flexible assemblies
	ASTM D1781-76(81)	Climbing drum test for skin-sandwich
	BS 5350: Part C13: 1980	Rubber-to-metal bonding
	ASTM D429-73	
Shear strength	ASTM D4027-81(86)	Modified rail test
	ASTM D229-70(81)	See Torque strength
	ASTM 2182-72(78)	Disk shear in compression
	BS 5350: Part C15: 1982	Bond strength in compressive shear
	BS 6319: Part 4: 1984	Slant shear test, loaded in compression, for resins used in construction; concrete substrates used
	BS5350: Part G2: 1987	Collar and pin bonded with anaerobic adhesive and loaded in tension

Table 4.3. Continued

Test	Standard	Remarks
Cleavage strength	BS 5350: Part C1: 1986	⎫ Compact tension specimen
	ASTM D1062-72(83)	⎭
	ASTM D3433-75(85)	Parallel- or tapered double-cantilever-beam joint for determining the adhesive fracture energy, G_{IC}
	ASTM D3762-79(83)	Wedge cleavage test (for aluminium adherends)
Fatigue strength	ASTM D3166-73(79)	Single-lap joint loaded in tension
Flexural strength	ASTM D1184-69(86)	Laminated assemblies
Torque strength	ASTM E229-70(81)	For determining pure shear strength and shear modulus of structural adhesives (napkin-ring specimen)
	ASTM D3658-78(84)	Specificallly for ultra-violet light-cured glass–metal joints
	BS 5350: Part G1: 1987	Anaerobic adhesives on threaded fasteners

Property	Standard references	Description
Impact resistance	ASTM D3807-79(84)	Plastics-to-plastics joints
	ASTM D950-82	Block shear specimen
	BS 5350: Part C4: 1986	
Creep resistance	BS 5350: Part C7: 1976	Various test geometries permitted
	ASTM D1780-72(83) and	Single-lap joint loaded in tension
	ASTM D2294-69(80)	
	ASTM D2293-69(80)	Single-lap joint, having long overlap, and loaded in compression
Environmental resistance	ASTM D2918-71(81)	Subjected to stress, moisture and temperature; uses peel joint
	ASTM D2919-84	As above, but uses single-lap shear joint loaded in tension
	ASTM D3762-79(83)	As above but uses a wedge test
	ASTM D1151-84	Exposure to moisture and temperature
	ASTM D1183-70(81)	Exposure to cyclic laboratory ageing conditions
	ASTM D904-57(82)	Exposure to artifical and natural light
	ASTM D896-84	Exposure to chemical reagents
	ASTM D3632-77(82)	Exposure to oxygen
	ASTM D1828-70(81)	Natural weathering
	ASTM D1879-70(81)	Exposure to high-energy irradiation

Durability evaluation requirements

The complete characterisation of an adhesive requires that its response to various load, time and environmental conditions be determined. Renton(32) says that useful data are:

(1) stress–strain response (in shear and tension) at a constant strain rate to failure
(2) cyclic stress–strain response versus the number of cycles to failure
(3) creep response
(4) combined shear and tension response
(5) the reaction of the first three items to various moisture and temperature environments.

Schematic representations of the typical response of cold-cure epoxies are depicted in Fig. 4.6. Naturally the relative importance of specific adhesive property data depends upon the application and the envisaged loading and environmental conditions that the real joint will be subjected to. Many analysts(32–34) advise that the 'strength' of the adhesive equilibrated with the 'worst-case' environment is the key to effective design. This implies laboratory tests conducted at high temperatures on specimens pre-equilibrated with high levels of water vapour or liquid water. For the application of adhesives to steel bridges in the USA, Albrecht *et al.*(34) selected a test environment of 49 °C and 90% r.h.

Test joints versus real joints

Real joints do not consist of simple, separate, elastic materials with a clear, mathematical geometry. The adherend surface is usually rough, and the thickness and properties of the primer (if applied) and adhesive layer are often difficult to regulate and to determine. There also exists some debate as to whether the in-bondline, or thin-film form, properties of the adhesive are the same as they are in bulk. In service, the applied load will in general be much lower than that applied in the laboratory, thereby imparting an 'acceleration' factor to the testing. It is important therefore to try to establish the relation between test life and real life.

Claims are made for the superiority of different specimen geometries or different test conditions, and there exists some

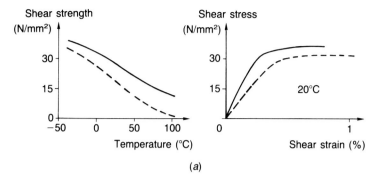

(a)

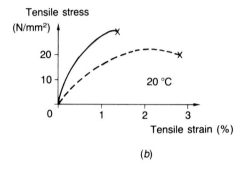

(b)

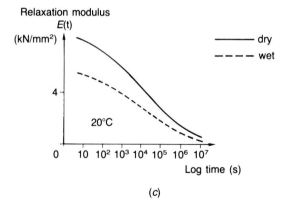

(c)

Fig. 4.6. Schematic representations of the generalised mechanical behaviour of cold-cured epoxy adhesive materials. (a) Shear response. (b) Tensile response. (c) Relaxation modulus from constant stress tests.

controversy world-wide over appropriate accelerated ageing procedures for bonded joints. This stems in part from the different objectives of the investigators, their backgrounds, and the different materials and processes involved. Formerly various lap shear joints, tensile butt-joints, and peel specimens were used. More recently fracture mechanics principles have been employed in cleavage specimens, and self-stressed configurations are favoured for durability testing. There remains a need for the development of appropriate, cheap, simple and quantitative test methods in order that adhesive materials may be brought more readily into the compass of general engineering applications. The test coupon approach to quality control is, perhaps, most constructively criticised by Albericci(35) with respect to aerospace applications of adhesives. The observation is made that problems which are likely to arise in the 'real' structure simply do not arise in the test coupon, and its validity tends to diminish with the complexity and size of the real component.

Joint tests in shear

Lap shear. The lap shear joint is that used almost universally in testing adhesives or surface preparation techniques. It owes its popularity to its convenience of manufacture and test, as well as to the fact that the adhesive is subjected to cleavage as well as to shear. It thus simulates, in a way that torsional shear test does not, the actual use of an adhesive. Its deficiency lies in the fact that the particular ratio of normal to shear stresses is likely to be very different from those situations in which it will be used, and which the test is intended to simulate. Thus, depending on such factors as adherend stiffness, overlap length, adhesive modulus, etc., failure of the 'shear' joint can be dominated by either shear or tension. Some results comparing the effect of different joint geometries (see Fig. 4.7) on the apparent shear strength of two cold-cure expoxies are collected in Table 4.4. Clearly, for joints constructed with thin adherends and/or long overlap lengths it makes more sense to quote joint failure loads and to describe the joint geometry, rather than to quote the average shear stress at failure.

The traditional view of lap joint failure is old and familiar and was, in essence, employed by Fairbairn(36) who considered the failure mode of riveted joints on iron bridges. However, the topic of bonded lap shear joints is probably best reviewed in detail by

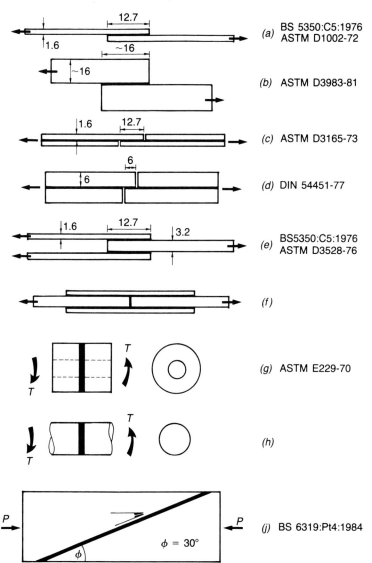

Fig. 4.7. Joint tests for shear.

Table 4.4. *Effect of joint geometry on the apparent shear strengths of two epoxy adhesives cured at 20 °C*

Joint type	Description/ standard	Overlap length (mm)	Adherend thickness (mm)	Average shear stress at failure (MN m^{-2})	
				Adhesive A	Adhesive B
Single lap	BS 5350:C5:1976	12.5	1.5	—	—
	ASTM D1002:1972	12.7	1.6	14–16	14–15
	(Fig. 4.7(*a*))	5.0	1.6	24–31	19–22
Double lap 'tuning fork'	BS 5350:C5:1976	12.5	1.5	—	—
	ASTM D3528:1976	12.6	3.2/1.6	15–17	18
	(Fig. 4.7(*e*))				
Cyl butt joint in shear	(Fig. 4.7(*h*))	25.4 dia	—	26	19

Doublelap	Ref. 87 (& Fig. 4.7(f))	40 eff	5.0	10–12(A)	18–21(C)
Single lap thick adherend shear test (TAST)	Similar to DIN54451:1977 (Fig. 4.7(d))	12.7	6.35	28(A)	27(C)
		10.0	6.35	33(A)	28(C)
		8.0	6.35	33(A)	27(C)
		8.0	6.35	40*(C)	27*(C)
		5.0	6.35	33(A)	26(C)
All-adhesive beam in shear box	Ref. 87 (& Fig. 2.15)	—	—	25–30	15–30

Notes:
Gritblasted steel adherends (* indicates silane-primed)
Joints cured and tested at about 20 °C
Bondline thicknesses of the order of 0.65 mm
Lap joint tests conducted to yield a rate of shear strain of about 1.0/minute, except for Ref. 87 double lap specimen
A adhesive failure
C cohesive failure
Adhesive A = aromatic amine-cured epoxy
Adhesive B = aliphatic amine-cured expoxy

Wake(37) and by Adams and Wake(5), and Kinloch(4) summarises
the evolution of the approach of the many stress analysts. The most
common shear test comprises the single lap shear joint embodied
in BS 5350(10) and ASTM D1002-72(11) (Fig. 4.7(a)). With reference to Figs. 4.1(a) and 4.8 it can be seen that the resulting stress
concentrations can be extremely large at the joint ends (points X
and Y in Fig. 4.8(b)). The analysis of Volkersen(15) predicts that
for identical adherends, the elastic shear stress concentration factor,
n, for the adhesive due to adherend tensile strain is given by

$$n = \frac{\tau_{max}}{\tau_{av}} = (\Delta/2)^{\frac{1}{2}} \coth (\Delta/2)^{\frac{1}{2}} \tag{4.6}$$

where $\Delta = \dfrac{G_a l^2}{Eht}$, the symbols being defined in Fig. 4.8.

Thus decreasing the overlap length or shear modulus of the adhesive,
or increasing the adherend stiffness or adhesive layer thickness, will
decrease the shear stress concentration in the adhesive layer.

The situation examined by Volkersen is incomplete in that no
account is taken of the tearing stress set up in the adhesive as a
result of the eccentricity of loading of the lap joint. The loads in
the single lap are not colinear so that a bending moment must exist,
and the joint will rotate as shown in Fig. 4.9. Goland and
Reissner(16) took this effect into account in their analysis by using
a bending moment factor, k, to account for cylindrical bending of

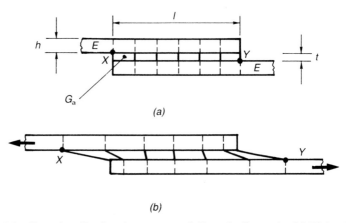

(a)

(b)

Fig. 4.8. Shear in adhesive due to extensibility of adherends. (a) Unloaded.
(b) Loaded in tension (elastic substrates).

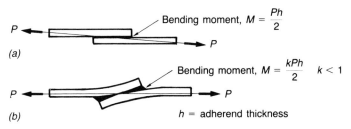

Fig. 4.9. Joint in bending (Ref. 5). (*a*) Undeformed joint. (*b*) Deformed joint.

the overlap region. Alignment tabs are sometimes attached to the ends of the joint to give a 'straight pull' during testing, but the joint will still rotate. Thus joints employing thin adherends or adherends in which the stress level is high will always bend so that the adhesive shear stress is non-uniform and large transverse peel stresses will exist. Average shear stress therefore bears little relationship to what is actually happening in such joints, and Lutz(38) describes some further effects of test specimen geometry on 'shear strength' results.

Attempts to improve on the single-lap test have led to a proliferation of alternative configurations (Fig. 4.7). The 'thick adherend' test (Figs. 4.7(*b*) and (*d*)) minimises the effect of differential straining by using stiff thick adherends and, with suitable extensometry, may be used to determine adhesive stress–strain response(39–43). The symmetry of the double lap joint (Fig. 4.7(*e*)) has led to the popular belief that bending is eliminated, but in fact it is really no more than a back-to-back arrangement of two single lap joints. Whilst there is no gross joint rotation under load, the internal joint loads are not colinear so that differential adherend straining and transverse stresses still exist. The double butt-strap joint shown in Fig. 4.7(*f*) suffers also from the same limitations.

Real bonded joints are unavoidably formed with a fillet of adhesive spew at the overlap ends (Fig. 4.10(*a*)). Even if this is removed, some slight amount of adhesive still remains and truly sharp corners are not encountered. It is in these regions of maximum stress, where failure is initiated, that most analyses are remote from reality, and here Adams and co-workers at Bristol University have been amongst the main investigators, using finite-element techniques(5). Indeed the adherend corner is likely to be rounded rather than truly square, which actually has the effect of reducing stress concentrations in the adhesive layer. Joint failure begins in these spew fillets of adhesive

145

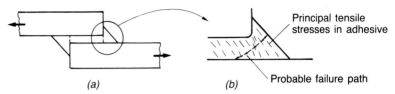

Fig. 4.10. Lap joint formed with spew fillet.

which, although not transmitting much stress, are subjected to large strains if adherend deformation is significant (Fig. 4.10(*b*)). Thus lap joint failure may be initiated by high tensile stresses within the spew.

Torsional shear. Torsional shear testing(5, 43–45) gives the best estimate of fundamental shear strength. By applying equal and opposite torques, T, to 'napkin ring' or annular butt-joint specimens (Fig. 4.7(*g*) and (*h*)), the adhesive is stressed purely in shear and the maximum stress, τ_{max}, will be that at the outside radius. However, it should be noted that the presence of a spew fillet of adhesive around the joint leads to a significant stress concentration at the adherend/adhesive interface. Bryant and Dukes(46) found that there was a linear relationship between decreasing bondline thickness and increasing failure stress. Curiously, this observation of the critical effect of adhesive layer thickness is not mentioned by any other investigator except Stringer(47). Lin and Bell(48) described the test in some detail, with an optimised specimen geometry and method of deformation measurement. Renton(32) reviewed the work of a number of investigators with the object of determining accurate shear stress–strain representations of the adhesive layer response. Stringer describes adhesive material characterisation using 15 mm diameter solid aluminium butt-joints. He found that the strain to failure of many adhesives was greatly increased by reducing the bondline thickness to below, say, 0.2 mm. A very large adherend deformation was subtracted from the total 'twist', and bondline thickness measurements needed to be determined very accurately for calculating material properties. ASTM E 229-70(49) describes a method of testing and evaluating the shear modulus and shear strength of adhesives by the napkin-ring test.

Scarf shear. It was stated earlier that scarf joint specimens are the most efficient way of smoothing the transfer of stress and strain across a joint. The scarf joint (Fig. 4.7(*j*)) is advocated for the measurement of 'bond strength' to concrete substrates by BS 6319(50); the Standard recommends that the average shear stress on the bondline at failure be recorded. A concrete prism is loaded in compression, and the bondline is subjected to a mixture of shear and compression, the relative proportions of which depend upon the scarf angle, ɸ. BS 6319 recommends 30° (giving a shear to compression ratio of 1.732 : 1) although other investigators have used 45° (giving a ratio of 1 : 1). Eyre and Domone(51) conducted a series of tests in which the scarf angle was varied, and the results were plotted on a Mohr's circle of stress, from which the average shear stress at failure for ɸ = 0° could be extrapolated for different adhesives. Again, this value is not an intrinsic adhesive material property, and the results are very dependent upon such factors as the concrete surface rugosity along the scarf. Further, it is quite common for the concrete to fracture away from the bondline. As a shear test it must be disqualified, and as test of 'bond strength' or adhesion it must also be discounted.

Joint tests in tension

Despite the fact that adhesive joints are rarely designed to be loaded directly in tensile mode, tensile tests are common for 'evaluating' adhesives. The axially-loaded butt (or 'poker chip') joint geometry, as recommended by ASTM D897(52), is depicted in Fig. 4.11.

A close look at the stress state induced within this joint indicates clearly a complex interaction of strain, and there are many useful commentaries on the limitations of this test(4, 5, 25, 31). If the adherend and adhesive moduli are very different (e.g. the ratio of epoxy to steel moduli, $E/E_a \approx 40$), the axial strains in the adhesive will be about 40 times greater than those in the adherend with a similar ratio for the lateral (Poisson's) strains. Where the two materials join, this conflict is resolved by generating large interfacial radial shear stresses (Fig. 4.11(*c*)). Joint strength increases with a decrease in adhesive thickness, and in a thin bondline affected completely by adherend restraint a complex stress arises. The ratio of the applied stress to the strain across the adhesive is then defined as the apparent or constrained Young's modulus, $E_a'(5)$.

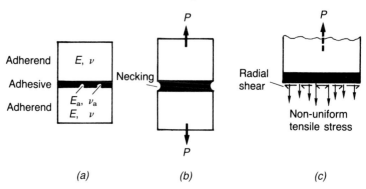

Fig. 4.11. Tensile (cylindrical) butt-joint test. (*a*) Unloaded. (*b*) Loaded. (*c*) Stress distribution.

If purely elastic behaviour is considered, elastic response of the adhesive may be obtained by attaching extensometers to the adherend either side of the bondline, making a suitable correction for adherend strain. Beyond the elastic limit, yield of the adhesive may be suppressed by the triaxial stress state, occurring at a stress higher than that in uniaxial tension. On the other hand, brittle adhesives which fracture before they yield will probably indicate a low failing stress because of stress concentrations.

One attempt to minimise the generation of large interfacial radial shear strains is reported by Renton(32), and advised by ESDU 81022(43). The geometry of the butt-joint is shown in Fig. 4.12 and it is implicit that the bondline thickness must be kept

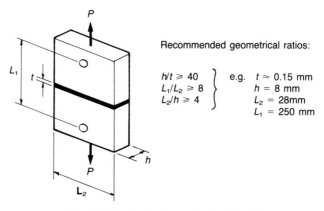

Recommended geometrical ratios:

$h/t \geqslant 40$
$L_1/L_2 \geqslant 8$
$L_2/h \geqslant 4$

e.g. $t \approx 0.15$ mm
$h = 8$ mm
$L_2 = 28$ mm
$L_1 = 250$ mm

Fig. 4.12. Tensile butt-joint (Ref. 32).

148

to a minimum if the overall specimen dimensions are to be kept within reasonable proportions. Renton states that this geometry gives a relatively uniform distribution of normal and shear stress over a large area, so that adhesive response may be determined from bondline deformation measurements. However, in the American evaluation programme this geometry was found to be very sensitive to temperature, moisture and strain rate. Moreover, deformation measurement was severely influenced by deformation of the aluminium adherends, and by included bondline air voids; a 10% void volume resulted in a 39% change in bulk modulus. For bondline thicknesses in excess of 0.6 mm, the specimen dimensions would clearly become very large.

It would seem that tensile stress–strain data determined from the butt-joint would be rather difficult to use to predict the failure of other than similar joints.

Joint tests for peel

Adhesives and adhesive bonds are very weak in peel, so that peel tests can discriminate rapidly between different surface pretreatments, particularly after, or during, environmental exposure (e.g. Brockmann(53)). Various forms of the peel test are described in several texts (5, 25, 37, 54–56), which are all essentially variations on the common theme depicted in Fig. 4.13, with the peel angle as the main variable. Adams and Wake(5) state that the key factor in determining fracture is the bending moment, M, at the tip of the propagating crack which is reacted over a very short length of adhesive, resulting in large local stresses, particularly in a direction across the adhesive thickness. However, the load measured by the peel test is not proportional to the 'strength' of the adhesive, but

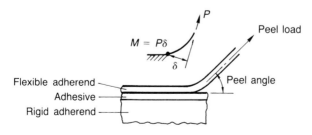

Fig. 4.13. Diagrammatic representation of the peel test (Ref. 5).

149

rather is influenced by its strain to failure, the bondline thickness, and the stiffness of the peeling adherends. The damped sine-wave pattern of stresses in a T-peel joint was illustrated in Fig. 4.1(d), from the work of Kaelble. Clearly the pattern of stresses is quite complex, but Adams and Wake showed that joint strength is predictable from bulk adhesive tensile properties, at least for a rubber-modified epoxy with a high strain to failure.

Joint tests for fracture

A major feature of the fracture mechanics argument is that the fracture energy, G_c, for a given joint, tested at a stated rate and temperature, is independent of the test geometry employed. In principle therefore, and with appropriate modifications, almost any test configuration could be used. In practice, certain geometries lend themselves particularly to analysis and experimental convenience, and are depicted in Fig. 4.14.

Mode I is the lowest energy fracture mode for isotropic materials and, thus, a crack always propagates along a path normal to the direction of maximum principal stresses. In joint fracture this is not necessarily the case since crack propagation is constrained to the adhesive layer, and mixed-mode effects may be important. Thus, attention must be given to joint fracture under additional loading modes for structural design purposes(28). Evidence in the literature suggests that the fracture energy of 'toughened' adhesives, when constrained as a layer in a joint, may be particularly sensitive to mixed-mode effects. Adhesive layer thickness may also have a profound effect because of its influence on the development of the plastic zone (Fig. 4.5). Bascom *et al.*(26) suggest that the size of this zone is very much larger than that associated with untoughened adhesives. Specimen width effects may also affect measured values because the state of stress varies from plane-stress in a very thin specimen to plane-strain near the centre of a wide plate. By conducting tests at a number of load rates and/or temperatures, time–temperature superposition principles may be employed to characterise fracture fully(25) in terms of both fracture energy, G_c, and fracture toughness, K_c.

Tapered double cantilever beam (TDCB). The TDCB, or more accurately the contoured DCB, specimen as developed by Mostovoy

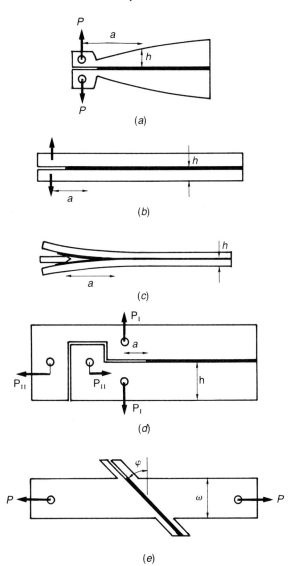

Fig. 4.14. Some adhesive joint fracture mechanics specimens. (*a*) Tapered double cantilever beam (TDCB). (*b*) Thick double cantilever beam (DCB). (*c*) Thin double cantilever beam or wedge cleavage specimen. (*d*) Independently loaded mixed-mode specimen (ILMMS). (*e*) Scarf joint.

and Ripling, is used frequently for ascertaining Mode I values (Fig. 4.14(a)). Provided that the arms of the specimen behave elastically, then the general expression for fracture energy is given by

$$G = \frac{P^2}{2b} \cdot \frac{\mathrm{d}C}{\mathrm{d}a} \tag{4.7}$$

where P is the applied load, b is the specimen width, and $\mathrm{d}C/\mathrm{d}a$ is the change in the compliance, C, of the structure with crack length, a. The explicit form of this equation for a TDCB specimen is given as

$$G_{\mathrm{IC}} = \frac{4P_{\mathrm{c}}^2}{b^2 E} \cdot m \tag{4.8}$$

where the constant

$$m = \frac{3a^2}{h^3} + \frac{1}{h} \tag{4.9}$$

h is the height of the beam at the respective crack length a, and m is often taken to be about 3.5 mm^{-1}. By tapering the substrates, a constant change in compliance with crack length is obtained, so that for a given applied load the value of G_{IC} is independent of crack length. This is particularly useful in experimental work since the location of the crack tip can be difficult to define accurately. A variant of this specimen design is one where the width of a parallel-double-cantilever-beam is increased down its length so that the specimen's compliance is again constant.

Load–displacement curves take the two major forms shown in Fig. 4.15. If the fracture energy is independent of the crack velocity, brittle crack propagation will occur at a constant load with the rate of propagation being dependent upon the strain rate (Fig. 4.15(a)); this is termed 'stable' crack growth. Alternatively, brittle crack propagation may occur intermittently in a stick-slip or 'unstable' manner (Fig. 4.15(b)), exhibiting load values appropriate to both crack initiation and crack arrest. This occurs when a crack, once initiated, extends spontaneously until the strain energy in the specimen is insufficient to continue propagation.

Major limitations of this test include the initial requirement for a numerically controlled milling machine to achieve the cubically curved specimen shape (for constant 'm'), the configuration is not

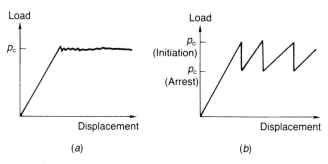

Fig. 4.15. Typical load–displacement curves for TDCB specimens. (*a*) Stable. (*b*) Unstable.

self-stressed, and bondline thickness is likely to affect the energy of fracture (especially with toughened adhesives).

Double cantilever beam (DCB). The 'thick' DCB specimen (Fig. 4.14(*b*)) is very popular in the aerospace industries for ascertaining Mode I values. The substrates are simple thick beams and inexpensive to produce. Joints can be loaded in a test machine, 'statically' or in fatigue, or else used as self-contained environmental exposure specimens when fitted with wedge-opening bolts(57–63). The specimen compliance, dC/da, is not constant so that under constant load conditions, G_I decreases as the crack length increases. This feature allows the establishment of 'threshold' levels as a function of environmental conditioning. Fracture energy is given by

$$G_{IC} = \frac{Ed^2h^3}{16} \cdot \frac{[3(a + 0.6h)^2 + h^2]}{[(a + 0.6h)^3 + ah^2]^2} \tag{4.10}$$

where: h = beam thickness; d = displacement at load point; a = crack length, from load point to crack tip; E = adherend modulus. The h^2 and ah^2 terms derive from the shear contribution to adherend elastic energy. The 0.6 h term is an empirical average correction for adherend rotation ahead of the crack tip, permitted by ductile strain or crazing of the adhesive(58, 63). The proper correction, unique to each adhesive, varies with the mechanical response to the environment. Crack lengths and locations (i.e. cohesive, interfacial, etc.) are monitored as a function of time of exposure with a travelling microscope, but this can be rather time-consuming.

The location of the true crack-tip is sometimes difficult to determine, even after careful filing of the specimen edges to remove adhesive spew. This can be particularly problematical if the adhesive is of coarse texture, when the crack is at the interface, or if significant adherend corrosion takes place during environmental exposure. Both sides of the specimen need to be monitored, and the average value taken. (In fact, crack propagation along the specimen sides tends to lag behind the advancing crack front in the centre of the specimen).

The behaviour of specimens constructed with either soft or stiff adhesives may be quite different(64). For instance, for the same value of '*d*' and '*a*' in each specimen, G_{IC} will be the same but the stiffer adhesive will be resisting a higher tensile stress. Clearly, creep effects with ductile adhesives will redistribute stress concentrations. There may also be a rate-dependent effect on G_{IC} (initial) when displacing the adherends with bolts.

Wedge cleavage test. A DCB derivative test developed by the Boeing Commercial Airplane Company is the wedge test or Boeing wedge test, standardised for aluminium adherends in ASTM D3762(65). The test specimen, depicted in Figure 4.14(*c*), may be regarded as a simplified, qualitative, and less expensive version of the DCB specimen. According to Marceau *et al.*(58) it was developed specifically to provide a simple specimen for adherend surface pretreatment process control for airframe construction. The method simulates in a qualitative manner the forces and effects on an adhesive bonded joint at the adherend/adhesive (primer) interface, and has proved to be correlatable with aircraft service performance. Essentially, a wedge is forced into the bondline and the stressed specimen is exposed to an aqueous environment. The resulting crack growth rate is monitored by travelling microscope and the locus of failure noted, i.e. cohesive, interfacial, etc. This test is probably the single most important tool used in the aircraft industry for monitoring and assessing adherend surface pretreatment.

The success of the test in discriminating between variations in adherend surface preparation and adhesive environmental durability has led to its widespread use in an R&D role(60, 66–69). In general, the test results are viewed only as qualitative because the fracture energies of many adhesives are high enough to cause inelastic deformation of the adherends. Thus crack lengths, rather than calculated fracture energies, are generally quoted. Nevertheless, Stone and Peet(63) demonstrated that an approximation to G_{IC} was a more useful basis for comparison than crack growth; indeed they conducted a very useful evaluation of the validity and limitations of the wedge test. If steel adherends are substituted for aluminium, adherend deformation should remain elastic for all but the toughest adhesives, and valid fracture energies may be ascertained from Eqn (4.10). The cleavage stresses induced at the crack tip are somewhat higher than in the stiffer DCB specimen because of the greater adherend deformation, and the test is likely to be more demanding of the interface. This greater deformation also facilitates location of the crack tip. Major limitations of the test are as for the DCB specimens.

Independently loaded mixed-mode specimen (ILMMS). This specimen is illustrated in Fig. 4.14(*d*) and, as the name suggests, shear and tensile loads are applied independently. The overall joint dimensions suggested by Bascom *et al.*(26, 27) of $300 \times 100 \times 12$ mm, imply oversize specimens; equations for the two fracture modes are

$$G_{IC} = (4P_I^2/b^2h^3E) \cdot [3(a + 0.6h)^2 + h^2] \tag{4.11}$$

$$G_{IIC} = [P_{II}^2/b^2h\,E] \tag{4.12}$$

The use of this specimen configuration is rarely reported, and some major limitations are that the large adherends require significant machining and are unlikely to be flat, the joint does not lend itself to stressed (durability) testing, the specimen width precludes rapid environmental ingress, and the monitoring of crack elongation in Mode I loading would be awkward.

Scarf joint. The importance of testing adhesive fracture under mixed-mode stress conditions has been noted. In the scarf joint (Fig. 4.14(*e*)), the applied load is resolved in the bondline to Mode I and Mode II components and their ratio changes with the scarf angle, ϕ. Bascom *et al.*(26, 70) investigated the effects of bondline thickness and test temperature, and $G_{(I, II)C}$ for $\phi = 45°$ was calculated from the failure load and crack length using a finite-element analysis. A complex behaviour pattern emerged, as discussed by the authors(26), and by Anderson *et al.*(25). It is very significant that the fracture energy of some toughened and some modified epoxies measured in combined shear and cleavage loading was lower than the corresponding Mode I fracture energy by as much as a factor of ten. Such vast differences were attributed to the effect of stress distribution on the morphology and micromechanics of failure. It should be noted that rubber modified products derive their toughness from a shear-yielding mechanism, so that mixed-mode loading may well interfere with this energy dissipative mechanism. Bascom and Hunston(27) concluded that in designing for flaw tolerance, Mode I fracture toughness is by no means a conservative estimate of fracture resistance – and may even be an unsafe assumption. Users of toughened adhesives beware!

This specimen configuration has the potential for highlighting some intriguing observations, but amongst its limitations must be that it is a massive specimen of complex shape requiring significant machining, it does not lend itself to stressed (durability) testing, and the bond area is too large to permit environmental access within a reasonable time-scale. Further, crack propagation will be focussed mechanically towards the interface in mixed-mode loading, involving surface roughness effects. Also, the scarf angle, bondline thickness, the nature of the adhesive, and the loading rate are all likely to affect the measured fracture energies.

4.5   Joint behaviour

Whilst the properties and behaviour of adhesives in bulk form are linkable to their composition, the behaviour of joints constructed with adhesives is less predictable. This is to be expected from a consideration of the factors affecting joint strength outlined earlier in the chapter. In particular, joint behaviour will be determined largely by the joint's geometrical configuration and by the way in which it is loaded, since these factors will determine the nature and magnitude of the resulting bondline stresses and stress concentrations. To a large extent, and provided that a reasonable amount of care in the design of the joint has been taken, the adhesive's stiffness will determine the general behaviour. However, because the stiffness of adhesives changes with loading and environmental conditions (e.g. time, temperature and moisture), consideration must be given to the effect of these conditions on joint behaviour.

Creep

The time-dependent component of polymer response is of extreme importance to the use of structural adhesives which are required to sustain either permanent or transient loads. At temperatures well below the adhesive's T_g, overloading is far more likely to lead to stress rupture than to creep(5). However, at temperatures close to or at T_g, some creep of loaded joints is to be expected. Recalling Chapter 2, highly cross-linked epoxies which are cured at elevated temperature possess the best resistance to creep.

Adhesives belong to the class of solids whose stress–strain behaviour may be described at visco-elastic. That is, in addition to having some of the characteristics of viscous liquids, they also possess some of the characteristics of elastic solids. Unlike an elastic material the strain lags behind the stress, so that it is necessary to describe the variation of stress and strain with time independently. The actual shear strain, measured as the increase in length of a standard lap-shear joint as it creeps, is minute. Allen and Shanahan(71, 72) studied the tensile creep of lap-shear joints at temperatures around that of the adhesive's T_g. They found that the creep under load was preceded by a delay or induction period which was temperature and load dependent. The steady state creep which took place was logarithmic with time, giving way eventually to an accelerated creep terminating in stress rupture. This behaviour was undoubtedly linkable both to the particular joint geometry and to the adherend and adhesive materials employed. One reason for the delay in the onset of creep may well be because of changes in the adhesive due to temperature and humidity, enabling some bondline stress redistribution – particularly if that results in a higher level of stress having to be borne by the central regions of a joint. Clearly in joint configurations comprising large bonded areas sandwiched between impermeable adherends, this delay will be very large indeed. Althof and Brockmann(73) advocated the measurement of bondline deformation in real joints, whilst subjected simultaneously to sustained loading and environmental exposure, to give limit values for design.

It is clear that there are load, and therefore stress, levels below which creep will not occur, but it must be recognised that changes in the adhesive's stiffness due to environmental conditions may well give rise to creep after a delay or induction period. This could, of course, be reversible such that steady state creep is not a necessary outcome. A better understanding of creep mechanisms provides a useful topic for further research. The practical approach to both delaying the onset of creep and to reducing its rate is by providing for large bonded areas – for example, long-overlap shear joints(19).

Fatigue

Fatigue and creep are intimately related, being facets of a common property of an adhesive. Thus, the adhesive's visco-elastic response is a major determinant of the fatigue life of the joint with which it is constructed. For most applications fatigue resistance in shear is of overriding importance, and slow-cycle loading is likely to be the most detrimental in allowing cumulative creep. Fatigue performance is again related to the joint configuration, and the resulting nature of the stress and strain distribution within the adhesive layer.

Romanko and Knauss(74) stated that an accurate representation of the adhesive stress–strain relation over the range of loads and environments expected in service is a prerequisite for understanding fatigue behaviour. They recommend monitoring of the load–deflection history of the adhesive layer in model joints, in the form of Fig. 4.16, of which the first few cycles of response are atypical. Romanko *et al.*(33, 74–76) monitored the creep strain by clip gauge, and described fully their data interpretation; Althof(42) monitored the strain with sophisticated extensometry. Krieger(77, 78) demonstrated that irreversible damage was caused by running a slow-cycle fatigue test in which the peak amplitude exceeded the adhesive's elastic limit. Marceau *et al.*(79) observed creep rupture at low frequency cycling of thick-adherend lap-shear joints.

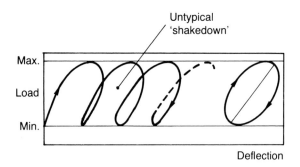

Fig. 4.16. Load–deflection history (Ref. 74).

The results of fatigue or Wöhler tests are customarily presented as *S–N* curves, in which the stress as a percentage (*S*) of the ultimate strength determined in a 'static' test is plotted against the number of cycles (*N*) at that stress to failure, on a logarithmic scale (Fig. 4.17). Matting and Draugelates(80) were the first to collect a large amount of data on adhesive joints of the single overlap configuration, and they showed that the survival time of joints from a large batch was distributed in a log normal distribution provided that the stresses involved were less than the so-called endurance limit. Above the endurance limit there will be a scatter about the Wöhler line, and below the survival time is indefinite (which does not mean that the joints are unbreakable!). Marceau et al.(79) confirmed that there was a frequency dependence, observing that the fracture modes of thick adherend lap shear joints changed, indicating creep-rupture at low frequencies. Allen *et al.*(81), in the culmination of many years' work, attempted to establish the reality of an endurance limit by combining stress analysis and fracture mechanics concepts to describe crack growth in the joints employed. Whilst enabling the calculations of critical crack lengths of flaw sizes, it was not established that cycling above the endurance limit induced crack propagation any more than cycling below it. Additionally, their approach must be tempered by two important facts, namely that purely elastic response to load was assumed and, secondly, that the cleavage stress calculated at the end of the joint was in error. The results of fatigue testing of double lap-shear joints constructed with cold-curing epoxies and steel adherends are reported by Mays *et al.*(82, 83).

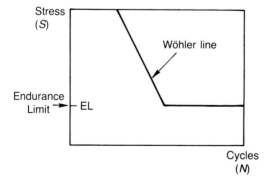

Fig. 4.17. Idealised Wöhler diagram (Ref. 37).

Fracture

Brittle fracture arising from crack formation and propagation applies more generally to joint failure rather than to joint behaviour. However, as indicated earlier, fracture energy (G_c) or fracture toughness (K_c) may be related to a critical crack length in a structure. Structures may therefore be designed to accommodate cracks of a pre-determined length and remain serviceable. It will be recalled that, like other material properties, G_c and K_c are not unique values. For instance, the plasticising action of elevated temperature and/or moisture absorbed by the adhesive is beneficial in subduing brittle fracture of the adhesive material.

Polymers tend to have rather lower fracture strengths than materials such as metals or ceramics, but not concrete! The theory of brittle fracture applies for polymers as for metals, but with greater emphasis on the development of a plastic zone around the tip of the growing crack (Fig. 4.5). The brittle mechanism is favoured in unmodified epoxies, and as a result of reducing the temperature, increasing the strain-rate or specimen thickness, and having sharp notches. Traditionally, susceptibility to brittle fracture has been assessed by some form of impact testing.

A high G_c determined from short-term tests does not guarantee resistance to long-term crack growth under transient or sustained loads, or in the presence of moisture. For example, crack growth occurring from fatigue loading may be plotted against G or K for different loading frequencies, in the characteristic form of Fig. 4.18. Comprehensive review material related to the fracture of bonded joints may be found in many texts (25–29), whilst Kinloch and Young(84) offer a monograph on the fracture behaviour of polymers.

Environmental conditions

It is apparent that the effects of moisture and heat, especially in combination with an applied stress, may have a considerable influence on bonded joint behaviour. This is because of changes in the stiffness of adhesive materials with exposure to environmental conditions, and the effect has been emphasised in previous sections.

The effects of temperature change only on joint strength fall into two categories. There is the general relation between the strength of a joint as a function of temperature, e.g. Fig. 4.19, and also as a

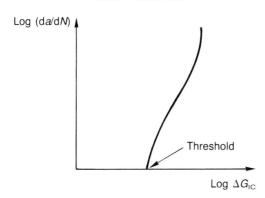

Fig. 4.18. Typical plot of da/dN versus applied fracture energy.

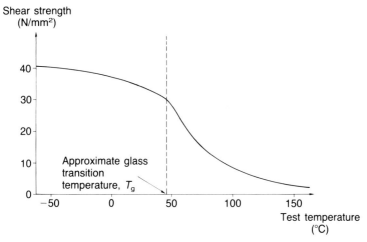

Fig. 4.19. Generalised representation of the mechanical strength of cold-cure epoxies as a function of temperature.

function of the time during which a given temperature is sustained. For instance, chemical changes in the adhesive may occur due to prolonged periods at elevated temperature (thermal ageing), or else there may be a time-dependent strain associated with stress relaxation or creep of loaded joints. Time–temperature superposition techniques as applied to polymers and pioneered by Ferry(85), are described fully by Hunston *et al.*(86). Lark and Mays(87) found that the cold-cure epoxies that they worked with were generally amenable to

161

these procedures in bulk flexure. Further, the fracture behaviour of toughened epoxies may also be described using these techniques(88).

4.6 Durability and performance in service

The durability of joints, and particularly structural adhesive joints, is generally more important than their initial strength. It has been found that the mechanical properties of a bonded joint may deteriorate upon exposure to its service environment and, further, that an interfacial locus of failure may often be found only after environmental attack. A number of empirical laboratory investigations reported in the literature many years ago established that water, in liquid or vapour form, is the most hostile environment for structural adhesive joints that is commonly encountered. The key factors affecting bond durability in general have been distilled from the literature(37, 69, 89–91), and Kinloch(69) brings together the most useful international contributions of adhesive scientists and technologists. The extent of the potential problem is illustrated in Figs. 3.1 and 4.20.

The problem with bonded joints is that most of the load is transmitted through the edge zones, and it is these which come under environmental attack first. In fact, the load becomes progressively borne by the inner region of the joint but, nevertheless,

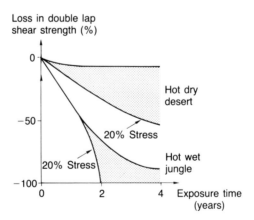

Fig. 4.20. Effect of outdoor weathering on joint strengths (Refs. 92, 93). Adhesive: epoxy polyamide film. Adherends: aluminium alloy. Pretreatment: Chromic-sulphuric acid etch. Cure temperature: 175 °C.

it is still the most highly stressed regions which are under the greatest environmental attack. It will be appreciated that the mechanical response of bonded joints is complex, so that the interpretation of the behaviour of environmentally exposed joints becomes even more onerous. For instance, the measured residual joint strength is a function of change both in the resin cohesive properties, and in the adhesion between adherend and adhesive. Thus joint durability demands at least a two-fold consideration of (a) the structural integrity of the cured adhesive, and (b) the environmental stability of the interface.

In so far as fracture loads are related to stress concentrations at the ends of a joint, then if the bonded area is sufficient to enable stress redistributions within the adhesive layer, changes in adhesive or cohesive properties will not compromise the integrity of bonded joints of an appropriate geometrical configuration. Heat alone is likely to be beneficial in post-curing the adhesive. The diffusion of moisture into the bondline may or may not be beneficial. The combined effect of heat and moisture however is likely to be detrimental, and the synergistic effects of heat, moisture and stress will certainly be detrimental. Some general observations on joint durability are summarised in Table 4.5, aspects of which are expanded in the following dialogue. It will become apparent that the environmental performance of bonded joints depends more on adhesive/substrate interfacial behaviour than on cohesive adhesive layer behaviour.

Effect of water

Water has proved to be the most harmful environment for bonded joints. Problems arise because water is universally found, and the polar groups which confer adhesive properties make the adhesives inherently hydrophilic; the substrates or substrate surfaces themselves may also be hydrophilic. Experience has demonstrated that the main processes involved in the deterioration of joints subjected to the influence of moisture are (a) absorption of water by the adhesive (b) adsorption of water at the interface through displacement of adhesive (c) corrosion or deterioration of the substrate surface.

It would be desirable to be able to predict the strength of joints exposed to their service environments from a consideration of the likely concentration and distribution of moisture within the adhesive

Table 4.5. *Parameters affecting environmental durability*

Water		Activity, bondline concentrations, pH and soluble aggressive ions. If absorbed by adhesive, may plasticise and toughen.
Temperature		Rate of degradation promoted by elevated temperature; also creep effects. May aid post-curing, and may plasticise and toughen cured adhesive.
Oxygen		Contribution to metallic corrosion and polymer degradation.
Adhesive	*rheology*	Interfacial contact. Air Voids.
	composition	Chemical type affects cured structure, bulk properties, interfacial composition and stability.
	cure schedule	Low temperature curing implies inferior performance.
Adherends	*metals*	Surface stability.
	concrete	Surface dryness and coherence. Permeability.
	composite	Moisture content. Permeability.
Surface pretreatment		A most important factor. Specific to particular adherends and, sometimes, the adhesive. Nature of primer (if applicable). Bonding conditions.
Stress	*internal*	Cure shrinkage, temperature variation, swelling by moisture.
	externally applied	Strained bonds more susceptible to attack. Probably increases the rate of diffusion of an ingressing medium. Stress-corrosion. Stressed polymer, with an increased free volume, may retain more water than an unstressed polymer.
Joint design		Stress concentrations/tensile forces at or near interface reveal sensitivity towards environmental attack.
Time		Duration of exposure, application of stress and adhesive viscoelasticity.

layer (as a function of time). Unfortunately this is not really possible(90), given the factors affecting joint 'strength' (Table 4.2). Nevertheless Brewis *et al.*(94) have observed a linear dependence of joint strength upon total water uptake for a number of bonded aluminium lap joints, and Althof(95) describes a technique for modelling the adhesive layer stress distribution as modified by moisture uptake. Such approaches cannot take into account the effect of moisture on the interface. As an example, the theoretical water concentration profiles within the adhesive layer of one quadrant of a lap shear joint, as a function of immersion time in water at 20°C, are presented three-dimensionally in Fig. 4.21. The highly filled epoxy adhesive was cold-cured with an aliphatic amine hardener.

Mechanisms of failure. Comyn(91) suggests that water may enter and affect the performance of a bonded joint by one or a combination of the following processes. Firstly, water may enter the joint by:

(1) diffusion through the adhesive
(2) capillary transmission along the adhesive/adherend interface (wicking)
(3) capillary action through cracks and crazes in the adhesive
(4) diffusion through the adherend if it is permeable (e.g. concrete, composite).

To the above may be added osmotic pressure gradients(96).

Secondly, having accessed a joint, water may cause weakening by one or a combination of the following actions:

(5) reversible alterations to polymer mechanical properties (e.g. plasticisation, swelling)
(6) induction of reversible bondline swelling stresses
(7) irreversible alterations to polymer mechanical properties (e.g. hydrolysis, cracking, crazing)
(8) irreversible adhesive/adherend interface attack, by displacing the adhesive or by hydrating a metal adherend oxide surface.

From the foregoing discussion interfacial attack(8) is clearly the most damaging action, but adhesive plasticisation also has a profound effect on joint performance.

Diffusion and absorption of water. Water is a remarkable substance, with special properties which can be related to its molecular structure, and which govern the way its molecules interact with each other and with other substances. The water molecule's polarity and ability to form hydrogen bonds makes it a universal solvent, allowing it to dissolve, soften or swell organic subtances whose molecules contain sufficient polar groups (especially –OH) such as epoxide(97).

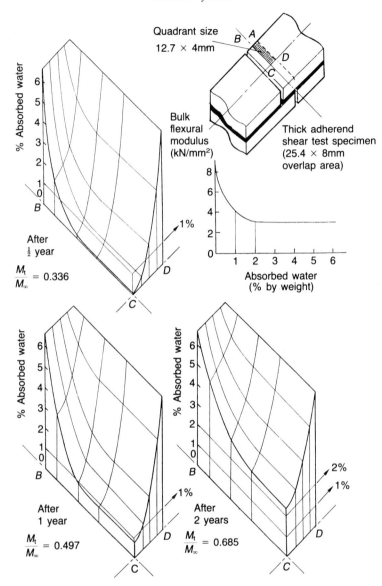

Fig. 4.21. Theoretical water concentrations within a lap joint bondline quadrant, after immersion in water at 20 °C. Adhesive: aliphatic amine cured epoxy at 20 °C. M_t = mass of water absorbed at time t. M_∞ = equilibrium mass water uptake at time t = ∞. Coefficient of diffusion D = 6.6×10^{-4} m²/s (H_2O, 20 °C).

166

Not only can the water molecule pass through gaps in the network because of its small size, but also because it is compatible with (i.e. effectively soluble in) the polymer. Thus polar adhesives are naturally hydrophilic whereas non-polar plastics, such as PVC and Polythene, are not; i.e. 'like permeates like'. The solubility of water in epoxies is of the order of a few mass per cent, and the coefficient of diffusion of water at 20 °C is around 10^{-13} m^2 s^{-1}.

For moisture to affect an adhesive joint between two metal adherends it must enter the joint by diffusion into the adhesive from an exposed edge. It is possible that moisture could 'wick' along the interface but this really implies either that the adhesive is displaced from the substrate at the exposed edge, or that shrinkage away from the adherend occurs. Where substrate surface corrosion takes place, the corrosion products themselves may help to strain adjacent adhesive bonds or the oxide layer itself may continue to grow at the interface, so enhancing the possibility of wicking. However wicking, as a primary mode of entry, is most unlikely and entry is normally gained by diffusion through the exposed boundary surface.

Diffusion, and adhesive displacement if it occurs, progresses inwards from the edges of the joint, and the rate of water transport through the adhesive to the interface is governed by the permeabililty (the product of diffusion and solubility). It follows that high permeability can occur with an adhesive in which diffusion is high or in which water is relatively soluble, or both. The rate of diffusion of water in an adhesive is important if water can displace the adhesive from its substrate, or if there is appreciable solubility, for this determines its rate, pattern and concentration in a bondline. If appropriate adherend surface pretreatments have not been carried out so that initial adhesion is minimal, wicking may in fact become a dominant mode of water entry. Alternatively, joint substrate surface perimeter corrosion may initiate interfacial cleavage forces and rupture the remaining adhesive bonds.

Heat-cured epoxies are often based on aromatic amines, which are less reactive than aliphatic amines, but which are much more rigid molecules (Chapter 2). This rigidity reduces their level of molecular motion and hinders water diffusion. With either type of curing agent, complcte cure will lead to more cross-linking and reduced water transport. Certain fillers may also reduce permeability and water transport.

Effect of water on the adhesive. The influence of water on the adhesive is generally reversible, so that any deterioration in, say, mechanical properties is recovered upon drying. The extent of this influence depends upon the adhesive's composition. All polymers absorb greater quantities of water when above their T_g, so that rubbery materials tend to show greater water absorption than rigid adhesives. Interestingly, the key position in structural metal-to-metal adhesives for airframe construction is occupied by epoxy–nylon adhesives, some of which display water uptakes of the order of 14%.

Water uptake by polymers is accommodated largely by swelling. For uptakes of only a few mass per cent, volumetric swelling would be of a similar or lower order(98, 99), and barely measurable. Moiré fringe interferometry has been used to quantify the swelling stresses developed in a layer of adhesive upon exposure to water(100), and Comyn(90) describes some other work related to calculations of the stresses induced in bonded joints by water sorption.

Water depresses the T_g of adhesives. This is worrying particularly for cold-curing epoxides with typical transitions when dry in the range 40°C–50°C, and underlines the need to select adhesives whose T_gs do not drop substantially with water sorption. Several attempts have been made to relate the depression of T_g to current concepts of the glass transition(90, 91). The modulus and strength of the cured polymer matrix are also lowered by water-induced plasticisation(34, 41, 95, 101, 102), in a manner akin to the organic plasticisers often used to modify the mechanical properties of adhesives. Brewis *et al.*(94) showed that for one particular hydrophilic adhesive/aluminium shear lap joint system, the depression of T_g could be used as a shift factor to relate the strength/temperature curve of dry joints to ones with saturated bondlines.

The fracture toughness of adhesives, other than toughened variants, generally increases with absorbed water, because of greater plastic deformation and enhanced crack-tip blunting mechanisms within a plasticised matrix(4, 28). Cohesive strength may, however, sooner or later be reduced sufficiently to offset the increased toughness. The general implication is that the toughness benefit of unmodified or initially tough products such as the rubber-modified adhesives is negated through a loss of cohesive strength(103).

The failure of a water uptake plot for thin films of adhesive (Fig. 2.22) to reach equilibrium can be taken as a sign of a chemical reaction between water and adhesive. Comyn(91) describes a number of studies of water-induced ageing processes including hydrolysis, cracking and crazing from cyclic climatic exposure, and the effect of absorbed electrolytes. Further, the hydrolysis of stressed bonds within the matrix (from internal stresses in joints, or from externally applied stresses on joints) remains a possibility.

Effect of water on the interface. The most important factor in the long term durability of bonded joints is the stability of interfacial adhesion against moisture. The absorption of water and its transport to the interface with the adherend can lead to irreversible changes, such as adhesive displacement by moisture and corrosion. Ultimately, the area of bond supporting the load diminishes until it can no longer sustain it, and joint failure occurs. The conditions leading to adhesive displacement and/or corrosion involve the nature of the adhesive bond operating in given combinations of adhesive and adherend. Brockmann(104, 105) has emphasised that the structure of the cured adhesive adjacent to the substrate surface differs from that of the bulk, because of the influence of the surface morphology and chemistry on the initial wetting and adsorption of adhesive. The inference is that this (weak) boundary layer of adhesive may be less densely cross-linked and/or have a lower concentration of filler particles than that of the bulk, and may be more susceptible to hydrolytic destruction. The rate of interfacial transport of water could also be somewhat higher than that through the bulk.

Widespread evidence indicates that the locus of failure of joints alters under the influence of moisture, from cohesive within the adhesive layer to interfacial separation. Water, in fact, displaces the adhesive when secondary valency bonds exist and in doing so applies stress to the fewer number of chemisorbed bonds which may be present. Adams and Wake(5) state that if the adhesive/adherend bond results solely from simple physical adsorption (secondary bonds), then the relative energies from adsorption of water to substrate and adhesive compared with the energy of adsorption of adhesive to substrate determine the equilibrium situation. Adhesive displacement is consequent upon a film of water existing preferentially at the interface if access is gained through defects, if the adhesive absorbs more than a few mass per cent of water, and if a high energy substrate surface is present. If the adhesive/adherend bond involves chemisorption, then displacement can only occur after hydrolytic destruction of the chemical bond.

The intrinsic stability of the interface in the presence of moisture may be assessed theoretically from the thermodynamic arguments advanced by Gledhill and Kinloch(106). Recalling the section of interfacial contact and intrinsic adhesion in Chapter 3, it is possible to deduce the thermodynamic work of adhesion, W_a, for different combinations of adhesive and substrate materials. Gledhill and Kinloch extended this to the calculation of the work of adhesion in a liquid environment, and showed that this value was negative for an epoxy/ferric oxide interface. Kinloch(69) and Comyn(91) present further predictions that bonds between epoxies and glass, aluminium, and iron and steel are unstable in water, whilst bonds to carbon-fibre composites are stable. Thus bonds to high-energy polar adherends are calculated to be unstable in the presence of moisture, and this is borne out by experience.

Joint strength rarely falls to zero but, rather, is generally lowered only. This may be because the adhesive layer has not absorbed sufficient quantities of water. It must also be recognised that the theoretical calculations assume secondary bonding only, and that chemisorption, mechanical interlocking and interdiffusion mechanisms of adhesion are not accounted for. The importance of establishing interfacial primary bonds through the use of primers and coupling agents was discussed in Chapter 3. The significance of micro-mechanical interlocking in contributing to joint durability was also discussed in that chapter, with particular regard to the micro-morphology and porous oxide structures of certain metallic substrate surfaces conferred by certain pretreatments. Oxide stability was emphasised.

Other possible interfacial degradative mechanisms include the build up of osmotic pressure at the oxide/adhesive interface(6) (akin to the phenomenon of paint blistering by osmotic gradients), disbonding by alkali produced by the cathodic reaction in metallic corrosion(107), and the imposition of stress leading to bond stress-corrosion cracking(108–110).

Effect of water and salt on the adherend. Deterioration of materials such as metals and concrete is often more rapid with salt solution than with water, for example by the action of electro-chemical corrosion. Water itself may be responsible for a number of changes in the adherends; concrete is likely to get stronger with further hydration of the cement, and plastic may become weaker by plasticisation. The resin/fibre interface in composite materials is also susceptible to degradation by water. With metals, water may attack the oxide layer and/or be involved in corrosion of the bulk adherend. Stress-corrosion of adherends, especially of thin section size, needs to be considered in tandem with fatigue resistance of bonded assemblies operating under relatively high stress levels. The fatigue resistance of metals is reduced dramatically through corrosion, by loss of section or by transgranular cracking in alloyed metals.

Techniques for increasing interfacial stability

Approaches made towards improving interfacial stability were, and to a certain extent still are, more empirical than scientific. This is because the exact nature of the mechanisms by which water disrupts adhesion is unclear which, in turn, reflects a lack of knowledge about metal/adhesive bond interactions. What is clear, is that the factors affecting intrinsic adhesion and bond strength degradation are different for different combinations of adhesive and adherend surfaces. This latter observation partially explains the former statements. Thus, two areas of interest emerge, namely to employ (a) hydrophobic adhesives, with suitable composition and rheology to ensure adherend surface wetting, and (b) appropriate adherend surface pretreatment.

The composition and rheology of adhesives was discussed in Chapter 2. The rate of diffusion and solubility of water in an adhesive is important if water can displace the adhesive from its substrate. Thus adhesive formulations which are essentially hydrophobic are desirable. This may be achieved by the use of highly cross-linked polymers, polymers containing plate-like fillers, the incorporation of hydrophobic additives, and polymers possessing only enough polar groups for adhesion. Bolger(111) discussed the balance of an optimum concentration of polar groups with polymer mobility. Mastronardi *et al.*(112) showed how the presence of hydrophobic coal-tar blended with the epoxy polymer matrix of coatings strongly reduced the affinity with water, minimising water transport and maximising durability. Bowditch and Stannard(113) describe an inherently hydrophobic cold-curing epoxy for steel bonding. Coupling agents may also be mixed into the adhesive.

An enormous literature is devoted to surface pretreatments for durable bonding, as discussed in Chapter 3. Suffice it to say that chemical etching, surface conversion coatings, coupling agents and primers all have a place among the techniques. It would seem that the lowering of the surface free energy of prepared metal surfaces by the deposition of coupling agents on, or coating of, their surfaces is very important because the adsorption of water becomes unfavourable. The possibility of establishing interfacial primary bonds (chemi-sorption) of these materials to the adherends is very interesting(69, 89, 91). For mild steel, gritblasting followed by the application of certain silane primers results in greatly enhanced durability(114, 115), (see Chapter 3). Other substances which can react with metal oxides and the adhesive to form water-stable bonds were also discussed in the previous chapter. The use of priming layers should be considered carefully as they may then themselves constitute the weakest link in the joint; cohesive failure may then occur within the primer layer.

The use of sealants to coat the edges of exposed joints has some merit in constituting a barrier to liquid water, but not water vapour. Most sealants are in fact more permeable than epoxy adhesives, so that a very thick layer would have to be applied in order to be effective. The usual joint-edge adhesive spew provides a useful barrier, as well as reducing stress concentrations.

Climatic exposure trials

It is very sound policy to collect and examine information on joints exposed to natural weathering conditions, rather than to depend solely on laboratory experiments. It has, however, been emphasised that because so many factors can affect joint strengths, extreme care must be taken when interpreting published performance data – and particularly data from durability trials. Nevertheless, the comparison of the results of outdoor exposure to tropical and temperate climates with those from laboratory testing would be expected to give a valuable indication of actual service behaviour(5). Such data have been few but the most complete account of trials organised by the Royal Aircraft Establishment was given by Cotter(93); some of their results are collected in Fig. 4.20. In Table 4.6 are summarised a number of durability studies, including outdoor exposure trials, each selected for their key findings and data interpretation, from a range of relevant experiences. Krieger(116) further advised that thirty years of airframe experience had not shown a failure in the cohesive mode which could be attributed to the environment; all problems had been at the interface.

In the early 1980s, the Wolfson Bridge Research Unit at Dundee University in Scotland embarked upon a programme for the environmental exposure of double lap-shear joints constructed with steel adherends united by cold-curing epoxies. The adhesive type and duration of exposure to each of seven environments were varied over a period of about 15 years. The joints were exposed in both an unstressed and in a stressed condition, and tested to destruction periodically both 'statically' and in fatigue. Rooftop exposure to the climate in Dundee (cool temperate) and Mauritius (hot humid) were included. All joints in the natural exposure environments fared worse than their counterparts exposed to laboratory-controlled conditions. Predictably, exposure to the hot/wet Mauritian climate resulted in the fastest and greatest degradation. The condition of the bonded interface of gritblasted joints after 2 years' exposure can be seen in Fig. 4.22, and classic joint perimeter corrosion may be observed. In fact, all residual joint 'strengths' were generally higher than the initial control values, with the notable exception of joints constructed with an epoxy–polysulphide adhesive. This fact emphasises the probability that the plasticising effect of imbibed moisture has made a positive contribution, through improved bondline stress redistribution for this particular joint configuration (of bonded area 80 mm × 25 mm wide). The data from natural exposure of course reflect the actual weather during the period of exposure, which may vary considerably from season to season.

Summary of durability aspects

The main durability aspects of bonded joints may now be summarised as follows:

(1) water is a particularly aggressive environment for bonded joints, especially when the bonded assembly is also subjected to conditions of relatively high (normal) stress and temperature

(2) joints comprising adherends possessing high surface free energies, e.g. metals, are particularly susceptible to environmental attack. Failure usually occurs at the adhesive/substrate interface

(3) the principal mechanisms of environmental failure identified are:

   (a) displacement of adhesive on the metal oxide by water, due

173

Table 4.6. *Some useful durability studies*

Reference (No)		Discipline/ application	Epoxide cold	hot	Adher. mat(s)	Pretreatments (followed by degreasing)	Adhesion promoters
Andrews & Stevenson (117)	UK	chem/academic		/	ti	polish	
Allen *et al.* (66)	UK	aerospace	/		al	various 'practical'	
Althof (41, 95)	Ger	DFVLR		/	al	chem etch	
Bascom (108)	USA	admiralty		/	al	chem etch	
Bethune (118)	USA	aerospace		/	al	chem etch	
Bodnar (119)	USA	various		/	al	various	
Brewis *et al.* (94, 120)	UK	chem/academic		/	al	various	
Brockmann (121)	Ger	DFVLR		/	al,st	various	
Cherry & Thompson (109, 110)	Aus	chem/eng	/		al	chem etch	
Comyn (90)	UK	chem/academic		/	al	various	
Cotter (93)	UK	aerospace		/	al,ti	chem etch	
Hockney (92)	UK	aerospace		/	al,ti	chem etch	
Wake (37)	UK	chem/academic		/	al	chem etch	
Krieger (64)	USA	aerospace		/	al	chem etch	
Minford (122, 123)	USA	aerospace	/	/	al	various	
McMillan (59, 60)	USA	aerospace		/	al	various	primers
Venables (124)	USA	aerospace		/	al	various	primers
Albrecht (34)	USA	civ eng/bridges	/	/	st		
Blight (125)	SA	civ eng	/		st,conc	?	
Bowditch & Stannard (113)	UK	admiralty	/		st,cfrp	grit blast	SPT
Brockmann (53)	Ger	DFVLR	/	/	st	grit blast	
Calder (126)	UK	civ eng/TRRL	/		st,conc	grit blast	some primed
Lloyd & Calder (127)	UK	civ eng/TRRL	/		st,conc	grit blast	
Mastronardi *et al.* (112)	I	chem/academic		/	st	sandblast	primers
Garnish (128)	UK	adhesive manufr.	/		st,s/st	various	
Gledhill & Kinloch (106)	UK	MOD/weapons		/	st	grit blast	
Gledhill *et al.* (129)	UK	MOD/weapons		/	st	grit blast	
Gettings & Kinloch (114)	UK	MOD/weapons		/	st	grit blast	silane
Gettings & Kinloch (130)	UK	MOD/weapons		/	s/st	chem etch	silane
Hutchinson (115)	UK	civ eng	/		st	various	silane
Kinloch & Shaw (28)	UK	MOD/weapons		/	st	various	silane
Jones & Swamy (131)	UK	civ eng/repair	/		st/conc		
Ladner & Weder (132)	Swi	civ eng/repair	/		st/conc	grit blast	primer
Nara & Gasparini (133)	USA	civ eng/bridges	/		st	grit blast	
Stevenson (134)	UK	chem/offshore			st	grit blast	primer
Sykes (107)	UK	phys/academic		/	st	various	
Trawinski (135)	USA	aerospace		/	st	cold chem etch	primers
Walker (136)	UK	AWRE/weapons	/		st	grit blast	silane

References grouped more or less
alphabetically in relation to adherend materials

al	aluminium & alloy
cfrp	carbon fibre reinf. plastic
conc	concrete
st	steel
s/st	stainless steel
ti	titanium

Table 4.6. *Some useful durability studies*

Joint type(s)	Const stress	Exposure environment	Fracture mechanics approach	Interface problem	Remarks
'blister'		water	/	/	
lap		water		/	evaluation of adhesives and pretreatments for aircraft repair
'thick' lap		water			bondline shear stress-strain response
various	/	water	some	/	REVIEW: stress corrosion
cleavage	/	water	/	/	test techniques
various	some	various	some	/	collection of papers
lap		water		/	water diffusion through adhesive
lap	/	water, outdoor		/	
'special'		water	/	/	effect of shrinkage stress
various		water		/	REVIEW: water diffusion
lap	some	outdoor		/	most extensive weathering trials
various	some	outdoor			ever reported and analysed for
lap	some	outdoor		/	temperate and tropical exposure
various	some	water	some	/	design data
lap	some	water, outdoor		/	includes tropical exposure
various	some	water, outdoor	some	/	REVIEWS: pretreatments and test methods
various	some	water	some	/	REVIEW: bond durability
various		some outdoor			work in progress
various	some	heat			creep, thermal props only
butt	some	water		/	'sacrificial' pretreatment technology
lap		water, outdoor		/	REVIEW: pretreatment
plated beams	some	outdoor		/	'strength' of ext. reinf. conc beams
		dye		/	microstructural investigation
coatings		water		/	coatings & corrosion
lap		water		/	test durations too short
butt		water		/	thermodynamic approach
butt		water	/	/	failure 'model'
butt		water		/	surface analysis
lap		water	/	/	surface analysis
cleavage	/	water	/	/	evaluation of adhesives
various	some	water	/	/	REVIEW: fracture mechanics
plated beams	some	water			in progress: 'strength' only?
plated beams	some			/	feasibility study
lap & butt	some	water	/	/	rubber/metal bond durability
'special'		water		/	REVIEW: pretreatment
lap		water	/	/	'new' pretreatment process
wedge		water		/	coatings & corrosion
'pulloff' & torque shear		outdoor			work in progress

water = liquid or vapour
different liquid compositions
constant or intermittent exposure
different temperatures

175

Adhesive joints

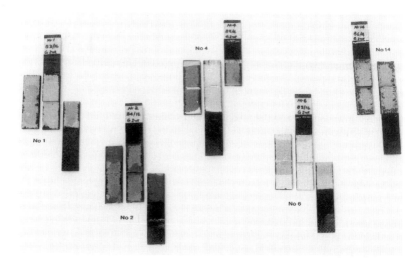

Exposure in Dundee

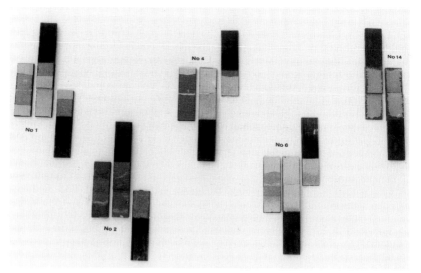

Exposure in Mauritius

Key to epoxide adhesive numbers	
No.	Hardener type
1	aliphatic amine
2	polyamine
4	aromatic amine
6	aliphatic amine
14	polysulphide

to the rupture of secondary bonds. This may be predicted from thermodynamic considerations

(b) loss of strength and failure of the metal oxide itself due to subtle changes, e.g. hydration

(c) hydrolysis in the boundary layer of adhesive adjacent to the adherend surface, the properties of this layer being different from those of the bulk adhesive

(d) the formation of interfacial pressure pockets by osmosis, if soluble salts are present on the substrate surface

(e) interfacial corrosion, which may or may not be a post-failure phenomenon

(4)   The kinetics of environmental failure are influenced by the diffusion of water through the adhesive. This may be altered by changing the adhesive's composition

(5)   appropriate surface pretreatments must be employed to create water-stable forces acting across the interface. The significance of this aspect is immense.

Information on the natural weathering behaviour of joints is very useful. By combining this information with data from 'accelerated' laboratory tests, some realistic predictions of service-lifetime may be made. Theoretical models of the pattern of bondline saturation of joints as a function of time of environmental exposure provide a useful appreciation of the possible extent of problems (e.g. Fig. 4.21). The process of joint failure, as observed in practice or in the laboratory, is frequently non-diagnostic; i.e. it rarely reveals the true cause, or the series of stages, leading to deterioration or failure.

4.7   Concluding remarks

The use of structural adhesives requires not only essential choices between the many types available, but also a considered decision on the design approach appropriate to structures assembled with them. The simplest way forward currently could involve the following steps:

Fig. 4.22. Condition of the bonded interface of unstressed double lap joints after 2 years' natural exposure. (Joints constructed with gritblasted steel adherends united by 5 types of epoxy adhesive, and cured at 20 °C).

177

(1) Adhesive selection and usage considerations from literature sources and/or proprietary software such as CADEPT (Harwell Laboratory), EASeL (Design Council), PAL (Permabond Adhesives), STICK (Lucas).

(2) Design with the aid of stress analysis programs (e.g. BISEPS, CADEPT, ESDU, PAL, PERA/CETIM, STICK), appropriate factors of safety and some reliability assessment method.

(3) Limited experimental work to investigate joint strength and durability.

Adhesives, as structural fasteners, are unique in that their stiffness can change with environmental conditions without any change in the loading on a structure which might be assembled with them. The adequate design of bonded joints must therefore take into account changes in the mechanical properties of adhesives, and the effect of such changes on the balance between the requirements of the different materials being joined. Rational design requires quantitative data on adhesives which requires, *inter alia*, appropriate test procedures for determining an envelope of their relevant structural properties. Probably the major deficiency in adhesive materials testing lies in the arbitrary test and report procedures adopted by the many investigators. Many test specimen configurations simply do not and cannot yield useful data, either to aid our understanding, or to be used for predicting performance in structural applications. In-bondline adhesive shear stress–strain response measurement is a useful method of characterising adhesives and of producing input data intended for structural analysis programs. However, bulk tensile response may yet prove to be a simple key element in design.

Proof of durability and safe performance are, rightly, onerous requirements for any innovations in the construction industry. The parameters affecting environmental durability have been summarised, and water has been identified as the most hostile environment for bonded joints that is commonly encountered. Identification of the general failure mechanisms is useful because it highlights the procedures necessary for the satisfactory fabrication of reliable and durable bonded joints. It also enables the development and adoption of appropriate test methods, since 'real' joint configurations are of limited use in assessing experimentally environmental effects (e.g. bonded areas must be minimised in order to allow environmental access within a reasonable time-scale). Fracture mechanics methods, especially those employing self-stressed cleavage joints, can provide an excellent means of examining adhesion rapidly, estimating bond durability, and for yielding data for direct use in design. Long-term joint durability is greatly dependent upon the stability of the adhesive–adherend interface. It follows that surface pretreatment is critical, and this is likely to be the most difficult aspect to control, particularly on site.

Real applications demand quality assurance in the form of appropriate physical and mechanical test procedures carried out alongside the actual fabrication. Naturally they should reflect the nature of the processes involved rather than, as in the ubiquitous concrete cube test, something almost totally unrelated. The value of a reliable method for assessing the post-bonding quality of bonded joints in the context of civil engineering applications will be self-evident. The application of non-destructive tests to bonded joints in general has been reviewed elsewhere(5). An interim solution which would at least provide the opportunity for strength tests would be to design joints from which test coupons could be removed, perhaps with a view to the long-term as well as to the short.

CHAPTER FIVE

Specification, fabrication and quality control

5.1 Introduction

The importance of quality assurance in any industry is widely acknowledged. More stringent customer expectations with regard to quality, together with the realisation that continued improvements in quality are often necessary for a company to sustain good economic performance, mean that some level of quality assurance is mandatory for all those associated with adhesive bonding operations. The use of technical specifications and quality control procedures go some way towards fulfilling the requirements of a quality system, and these measures are outlined here.

Most of the aspects relating to the selection and use of adhesives have been discussed in the first four chapters. For example, the properties of adhesive materials have been reviewed, together with the design approach appropriate to structures assembled with them. The importance of surface pretreatment and its influence on joint durability has also been covered. However adhesion science must also interface with fabrication considerations and with structural engineering. To the technical difficulties facing the production engineer in using structural adhesives must be added a number of others for the civil engineer. Primarily these are that:

(1)   each project tends to be a 'one-off' production
(2)   fabrication commonly takes place on site and so the environment is not easily controlled
(3)   the success of the bonding operation is very dependent on the diligence employed by the operatives and the work requires skilled supervision
(4)   the bonded assembly has to last for many decades
(5)   adhesive polymers have poor fire resistance

A number of general fabrication precepts applicable to civil engineering uses of adhesives have been collected in Table 5.1.

Table 5.1. *Example fabrication precepts*

(1) trained operatives must be employed under skilled supervision

(2) surface pretreatment and cleanliness are essential for a reliable and durable joint

(3) apply adhesive to prepared surface as soon as possible after surface pretreatment

(4) safe handling precautions are essential for resin products

(5) adhesive products should be stored in accordance with their manufacturers' recommendations

(6) thorough mixing of correctly proportioned components is essential; avoid sub-dividing manufacturers' pre-weighed packs; proper dosing equipment is desirable

(7) mix cold-cure products in quantities of less than 10 kg to reduce the exotherm in the mixing vessel and to minimise air inclusion; encourage the dissipation of heat from the mixed components

(8) ensure sufficient usable (pot) life with reference to 7

(9) match the reactivity of the adhesive to the temperature and duration of application; note that most adhesives will not cure below 5 °C

(10) superior performance is obtained from elevated temperature curing; as a rule of thumb, the reaction rate doubles and the cure-time halves for every 8 °C rise in temperature

(11) ensure gap-filling properties

(12) employ thixotropic or pseudo-plastic materials for application to vertical surfaces

(13) devise a suitable method of adhesive application to the substrate surfaces

(14) provide jigs/temporary works/permanent mechanical fixings to hold components and to apply pressure during cure

(15) adherend temperature should preferably be higher than ambient temperature to minimise condensation of moisture at the interface before adhesive application

(16) close the joint as soon as possible after application of the adhesive, and in a manner and at a rate that minimises the inclusion of air

(17) provide for heat-curing if necessary with heating tapes, focussed infra-red heaters, etc.

(18) provide appropriate physical and mechanical tests for quality assurance

(19) protect bondline from adverse environmental effects at least while curing

(20) consider post-bonding quality assurance by appropriate methods, including NDT.

Table 5.2. *Check-list of adhesive selection considerations*

Strength requirements	
Shear	unlikely to be limiting, but if so use a suitably modified epoxy
Cleavage/tension	sacrifice shear strength for high toughness and good peel strength with flexibilised or toughened epoxy
Impact	high toughness derived from toughened epoxy; good peel resistance conferred by flexible materials such as epoxy–polysulphide formulations
Deformation characteristics	
Modulus	high for unmodified epoxy
Creep	use adhesive with T_g well above service temperature; best resistance with unmodified epoxides
Service conditions	
Moisture, temperature	best resistance with unmodified epoxides and formulations with high Tgs
Stress	modified, flexibilised and toughened grades to suit function

5.2 Specification considerations

An example 'Compliance Spectrum' for steel/concrete bonding was proposed by the authors some years ago(1), and this is included as an Appendix to this Book. Elements of this are incorporated within the discussion on steel/concrete bonding which may be found towards the end of Chapter 6.

Elementary concepts

The important factors involved in any specification relating to the formation of bonded joints must include a consideration of:

(1) adhesive selection
(2) surface pretreatment
(3) joint design
(4) controlled fabrication procedures

Adhesive selection

There exist a number of sources listing the general factors involved in adhesive selection and the performance properties to consider (e.g. 2–5). It has already been emphasised that the adhesive represents only one part in a joining process which involves, among other things, a consideration of the needs of the various materials in the bonded assembly. For example, thin sheet material and polymer composite materials require either a toughened product or a flexibilised adhesive with high strain to failure in order to accommodate gross adherend strain under load. Similarly concrete adherends may benefit from being bonded with flexible and relatively low modulus products in order to reduce interfacial stress concentrations which may initiate surface fracture of the concrete. On the other hand, sustained loading may lead to excessive deformation and creep unless the adhesive is relatively unmodifed and therefore highly cross-linked, whilst also possessing a high T_g. A summary check-list of adhesive considerations is offered in Table 5.2; clearly, there exist property trade-offs.

Obviously some adhesive mechanical property data must be specified together with a detailed elaboration of the test procedures used for obtaining them, preferably over a range of conditions.

Procedures for the determination of important properties are essential for the selection of an appropriate product and for joint design and analysis. We have seen that adhesive polymers are sensitive to temperature and moisture, and to the rate at which stress is applied. Their bondline behaviour in joints depends greatly upon the system stiffness, and upon more subtle variations such as bondline thickness. It is therefore important that test methods are relevant to the real application, the fabrication conditions, the actual materials to be joined, and so on. Many standard test methods exist, for both strength and durability assessment, and these were discussed in Chapter 4.

For cold-curing epoxides wide variations in adhesive material properties are possible, with different combinations of resin, hardener, filler, and the multitude of modifiers. Products which cure at ambient temperature cannot achieve the same performance as is obtained by curing at elevated temperature. For products cured at room temperature their T_gs, at 40–50 °C initially, are relatively low and may be lowered even further by absorbed water, in liquid or vapour form. This may also be accompanied by a reduction in strength and modulus. Thus the use of materials with a slow and small water uptake is to be preferred, which implies a fairly highly cross-linked formulation. Such considerations do of course depend upon the performance and durability expectations in service. Whilst the environmental durability of joints can often be improved enormously by the surface pretreatment methods employed (see Chapters 3 and 4), the adhesive must be selected carefully to ensure long term durability in consideration of the modes and duration of loading, and the environmental conditions. Ideally the adhesive should be fairly tolerant of poor surface pretreatment procedures.

The working characteristics of the adhesive relevant to the application conditions must be determined. For instance viscosity is often temperature, shear-rate and time-dependent, and this will influence the choice of dispensing equipment, the method of application, the usable life and the open time. The viscosity should therefore be regulated bearing in mind the adherend rugosity and surface pretreatment, the method and location of application, and the cure temperature and duration of application. A thixotropic material may be required for application to vertical or soffit surfaces. Generally, relatively thick bondlines are encountered so that the adhesive should be able to cure in thick and/or uneven layers. It should also be remembered that for about every 8 °C change in

184

temperature, the reaction rate changes by a factor of two, so that the full cure time for an adhesive may vary widely depending upon ambient temperature; adequate strength may however be conferred within a fairly short time. Finally, it is usually important that the adhesive should undergo negligible shrinkage on cure (polyester-based adhesives can shrink substantially).

It is important that the adhesive manufacturer should be assessed for the quality and consistency of his products. Quality control test data should be available for each batch of adhesive, which requires the prior definition of the necessary characteristics and the means of assessing them. Adhesive materials should be packed in suitable containers ready for mixing, and each container should be durably and legibly marked with appropriate information. An instruction sheet should also accompany any delivery of adhesive, detailing information which includes product chemical type, storage conditions, mixing and application instructions, advice on compatible primers, curing profiles, safety instructions, and so on (e.g. see Appendix).

Surface pretreatment

It has been emphasised that the achievement of a high standard of surface pretreatment is the key to maximising joint durability and joint quality. Unfortunately, in the practice of adhesive bonding for applications in construction, surface pretreatment is likely to be the most difficult process to control. The choice and specification of pretreatment procedures should be influenced mainly by the required durability and, ideally, should entail simple reproducible processes. However the location and scale of operations, the nature of the adherends, the adhesive to be used, the safety and environmental aspects, and of course the cost, all have to be taken into consideration. The performance of joints constructed with cold-cure epoxies is likely to be critically dependent upon surface preparation. Some heat-cured products, notably the acrylics and epoxies, are claimed to be tolerant of oily and greasy surfaces, but their use should not be seen as an alternative to good surface preparation bearing in mind the usual civil engineering long term durability requirements. The stability of the adherend/adhesive interface is probably the most important factor in the durability of bonded joints.

185

Some materials can create particular problems for surface preparation and for obtaining adequate adhesion, and so should be avoided if possible. Bonds to metal alloys and plastics have been well researched, and a number of procedures are documented. Bonds to cementitious and siliceous surfaces are also readily obtainable, although little is really known about optimising surface treatment processes for maximum adhesion. Regard should be paid to the dusty, friable nature of concrete surfaces, as well as its high alkalinity.

For mild steel adherends, gritblasting is likely to be the minimum appropriate pretreatment. Careful specification of rust removal and degreasing procedures prior to blasting is necessary, as is the identification of the type of grit, pressure of blasting, and so on. For stainless steels, a non-ferrous grit such as alumina should be employed. Following blasting dust removal is desirable (e.g. by brushing or blowing), but further degreasing is not advised. The chemical etching of concrete is likely to lead to many problems unless the necessary procedures are carried out to the letter of a detailed specification; any chemical residues will cause long term durability problems at the adherend/adhesive interface.

The use of adhesive primers may be more important in some instances than others (e.g. on porous surfaces). Experience has shown that, despite the extra process involved, the advantages to be gained by priming in terms of presenting a reproducible surface to the adhesive far outweigh the possible disadvantages. Their use may also negate the need for complex surface pretreatment procedures. Certain siloxane coupling agents have been shown to provide a significant increase in joint durability, as have corrosion-inhibiting primers. Compatibility with the adhesive must of course be first established.

The control of surface pretreatment procedures may be done optically, by assessment of surface wettability, by surface analytical means, or by the use of simple mechanical test procedures. (See 'Adhesion control'.)

Joint design

The design of bonded joints must take into account the nature and magnitude of the likely operating stresses and strains, and should be contrived so that unacceptable peel or cleavage forces are not introduced into the bondline. Such design should also allow for changes in the mechanical properties of adhesives which may take place under different operating environments. As with all joints, they should also be designed for ease of location and assembly, and consideration should be given at the design stage to the means of clamping or holding of the component parts whilst the adhesive cures. The control of bondline thickness may also be achieved by making components self-locating or by the introduction of 'designed-in' spacers. Small changes to a joint's design may help very significantly in ensuring location and speed of fabrication, as well as with quality of the finished assembly.

As a further quality check from strength tests, it may be possible to design joints from which test coupons could be removed periodically for assessment.

Fabrication procedures

Probably the most critical aspect of the introduction of a new process is that of the education and training of the personnel involved. It is vital that the people have a qualitatively correct overall picture of the importance of the different stages of the bonding operation, and of the health and safety considerations. The use of an untrained workforce is likely to cause problems sooner rather than later!

Control of the working environment may be very difficult. Very low temperatures (below, say, 5 °C) or excessive humidity may cause problems with curing or with bond performance. General standards of cleanliness and surface pretreatment procedures must be well defined, but they do share commonality with those used in painting and corrosion protection. Variations from standard curing conditions may result in the adhesive not achieving full strength within a reasonable time-scale. Thus close attention should be paid to temperature, bearing in mind the thermal mass of the components involved; the need to specify heat-curing methods may be necessary. It is often desirable that the adherend temperature should not be lower than that of the surrounding air prior to application of the adhesive, in order to prevent moisture condensing at the interface.

The time elapsed between surface pretreatment and application of the adhesive should generally be kept as short as possible, in order to minimise subsequent contamination. By priming the adherend surfaces involved, either at the job-site or elsewhere, more reliable and reproducible surfaces will be bonded with a greater degree of implied quality.

Adhesive mixing, dispensing and application must be controlled carefully. Automatic dispensing is perhaps preferred although the use of adhesive packed in suitable containers, so that weighing or measurement is not required, is desirable. Application of sufficient, but not excess, adhesive is necessary and consideration of the means of so doing is desirable at an early stage. Insufficient adhesive will result in joint performance and durability problems, whilst a gross excess will hamper assembly and necessitate time-consuming clean-up operations. Joints should be closed in a manner and at a rate that minimises the inclusion of air, whilst with close fitting parts over large bonded areas a means of air escape should be provided. For example, holes may be drilled at regular intervals in large impermeable adherends such as steel plates, a procedure employed in some operations involving external steel plate reinforcement.

As mentioned previously a good joint design will assist the positioning and location of the joint parts, but jigs or permanent mechanical fixings may also be required to hold the components whilst the adhesive cures. Bondline thickness control will also be required, not only to give uniform design thickness but also to prevent displacement of the adhesive under pressure from clamps or jigs. It can be desirable (from a structural point of view) to leave fillets of adhesive surrounding the bonded joint, and these can be tooled appropriately once the joint has been closed.

Physical and mechanical quality control test procedures should ideally be carried out alongside the actual fabrication, and these are reviewed briefly below.

5.3 Quality control and non-destructive testing

Appropriate test methods for the control of fabrication procedures and non-destructive testing (NDT) are basic requirements for the formation of structural adhesive joints. Such methods should be based upon relatively easily measurable parameters that have a close identity with the properties of the bonded assembly that need to be controlled. However the quality of bonded joints depends upon many factors, requiring a range of very different procedures.

There are two essential but different quality aspects to be considered. The first is the adhesive strength of the bond between the polymer and the substrate, requiring control of the quality of that interface. The second is the quality of the adhesive strength of the polymeric layer, requiring control of the quality of the cured adhesive layer.

If the adhesive layer was always to represent the weakest link in the joint, the quality of the assembly would be limited to an assessment of the cohesive properties of that layer. Unfortunately it is the interface which tends to be of greater concern because it is here that joint failure is more commonly encountered, particularly after ageing. The most useful tests for indicating interfacial quality are in fact pre-bonding surface inspection procedures, which can nevertheless suggest potential joint performance. The tests used for assessing cohesive properties of the adhesive layer must necessarily be post-bonding activities.

Most of the NDT techniques which have been developed are associated with determining cohesion quality. However those techniques currently available have been developed largely for the aircraft industry, and are not generally applicable to the scale, location or materials involved in civil engineering structures. It is hoped that this situation will change in response to demand, and a brief review of some promising methods is included where appropriate.

Adhesion control

The measurement of the strength of adhesion between adherend and adhesive requires a measurement of the intermolecular forces of attraction; this is not currently possible. This aspect of quality control is therefore essentially reduced to assessing the adherend surface characteristics prior to bonding, although some post-bonding simple mechanical tests are appropriate.

Weak or loose surface layers are to be avoided, as are excessive amounts of water vapour, hydrocarbons or other surface contaminants.

Quality control activities must therefore address:

(1) removing weak or loosely bound surface layers
(2) reducing surface contamination to a minimum
(3) verifying the desired surface texture or oxide layer.

The methods used in addressing these activities may include:

Visual inspection. Simple optical methods and electron microscopy can be used for examining surfaces. The latter method can be used for assessing optimum oxide formation, where this is critical, or aluminium and metal alloys, as well as general morphological characteristics.

Wettability tests. Surface wettability may be readily assessed simply but subjectively by measurement of the contact angle. If the surface is clean, it is readily wetted and a drop of water will spread out rather than remain as a discrete droplet. This method cannot really be used to detect small variations in quality, but rather the gross effects. It does not lend itself to use on very rough or porous surfaces such as concrete.

Surface potential difference. Absorbed layers of contamination on metallic surfaces can be detected by measurement of variations in the amount of energy required for an electron to leave the surface. The Fokker Contamination Tester developed from this principle is claimed to be very useful in the optimization of various surface treatment processes and routine quality control activities (6).

Mechanical test procedures. As discussed in Chapter 4, mechanical tests of adhesion must introduce a tensile force at the adherend/adhesive interface. Peel tests can therefore discriminate rapidly between different surface pretreatments, or between surfaces carrying different amounts of surface contamination, particularly after environmental exposure. The same is true of the fracture mechanics test specimen configuration such as the wedge cleavage test (Fig. 4.14(c)). The Boeing wedge test was developed as a cheap and sensitive method of discriminating between variations in aluminium adherend surface preparation, and the concept has been extended to other metals and even other materials. Tensile pull-off tests have traditionally been used in connnection with concrete substrates, in which a steel or aluminium dolly is bonded to a prepared concrete surface. The torque required to pull the dolly off the surface can indicate something about the surface treatment only if failure is experienced at the interface – an outcome which rarely happens.

Cohesion control

Tests to monitor the quality of the adhesive materials may relate to the fresh state (e.g. viscosity, sag, pot-life), and to the hardened state – requiring measurement of mechanical properties.

Quality variations in the cured adhesive layer itself can occur as a result of a number of factors. Voids may be present due to air becoming trapped by the method of mixing, dispensing or application of adhesive to the substrate; small voids may be caused by volatiles. Surface voids may be present due to adhesive being applied to one adherend only before closing the joint. Local areas of uncured material may be present due to poor or incorrect mixing of two-component systems, or due simply to insufficient curing. Small cracks can appear in the adhesive layer due to shrinkage problems with cure, or of course due to large applied stresses when cured – particularly in a cold environment. In general, poor or insufficient cure is self-correcting with time inasmuch as chemical reaction will continue.

The presence of such defects may or may not be important, depending on their extent, nature, location and the operating stresses and environment to which a joint is subjected. For the majority of joints, the presence of relatively small defects is of little immediate significance unless they occur in regions of high stress transfer (e.g. at the ends of a bonded overlap). In the longer term such defects may allow faster accession of water or sites for fatigue crack nucleation, and as such it is important to detect them. The NDT methods available are claimed to give information on the following:

(1) density of the adhesive layer
(2) thickness of the adhesive layer
(3) interfacial debonds
(4) overall structural stiffness

Physical and mechanical tests. Three main sources of guidance which exist in the UK are:

(1) BS5350 – Methods of test for adhesives (7)
(2) BS6319 – Testing of resin compositions for use in construction (8)
(3) FIP – Proposal for a standard for acceptance tests and verification of epoxy bonding agents for segmental construction (9)

It may be useful to monitor the unreacted components of the adhesive system as well as the properties of the freshly mixed adhesive, partly as a check on the adhesive supplier and partly to establish the working characteristics of the adhesive. Techniques for measuring the viscosity of adhesives are defined in BS5350 : Part B8 : 1977, whilst aspects such as open time, pot-life and thixotropy (sag flow) may be established for the actual application. Suggestions relevant to the use of epoxy bonding agents for segmental construction are documented(9).

Tests for determining mechanical properties of the cured adhesive may take the form of the fabrication and testing of bulk adhesive specimens and adhesive joints. As discussed in chapter 4 the standard test procedures listed by ASTM, BSI, DIN and other official bodies are essentially for testing adhesives and surface treatments rather than joints. A number of methods therefore exist for the testing of bulk adhesive specimens – tension, shear, compression and thermal response properties being fairly easily determined. Interpretation of the results of joint tests must be done very carefully, requiring also a careful choice of test joint geometries. This topic was discussed fully in Chapter 4, and a list of standard test methods is given in Table 4.3. Some of the bulk and joint test procedures are elaborated in the Appendix. On the whole, joint tests are of limited value in assuring quality of the bonded assembly.

Non-destructive testing. NDT represents a large and diverse field in which a number of review papers have been published in recent years(6, 10–13). Brief mention only of the most important techniques is given below because few are currently readily applicable to adhesive joints in civil engineering. The techniques in general are void detectors.

Sonic methods The coin-tap technique is one of the oldest and still intuitively acceptable methods. The sound emitted by tapping the surface of well-bonded adherends differs from that over an unbonded area indicating voids, disbonds, and so on. Voids of two or three centimetres in diameter can be detected but it remains a very subjective test. Attempts have been made to quantify the result by exciting the bonded structure with an electromagnetic tapper, measuring its response and then analyzing the frequency spectrum(14, 15). Trials have been conducted on the experimental Wester Duntanlich Bridge which was fabricated using concrete-to-steel-bonded precast deck panels, and is discussed at the beginning of Chapter 8. Areas which were well-bonded gave a heavily damped response, while at small areas which were delaminated a less damped response was produced.

Ultrasonic methods There exist a variety of methods which are used widely in the aircraft industry. The technique measures changes in acoustic impedance caused by defects in a bonded assembly when an ultrasonic transducer is liquid-coupled to it. The high frequency sound waves are simply scattered by the presence of porosity or voids in the bondline. Interpretation of the data can be difficult, although substrate thickness is not generally a limitation.

Acoustic emission If a bond is mechanically or thermally stressed local perturbations of energy, or stress waves, may be released from discontinuities such as disbonds. The high frequency content of such stress waves may then be detected with a piezoelastic sensor. Unfortunately it is usually necessary to stress the joint to a considerable extent, which may often be impossible or inadvisable.

Thermography Heat-sensitive detectors may be used to ascertain the temperature distribution over a heated surface; disbonds will show up as hotter areas provided the adherends are relatively thin. Conversely, by heating one surface of a bonded sandwich structure and monitoring the opposite face, areas of disbond to which the heat has not been conducted will become apparent as cool spots.

Radiography Density variations in bonded components can be detected with X-rays, provided that metallic fillers or other special contrast indicators are present in the adhesive. X-ray inspection of bonded honeycomb sandwich panels may be used for establishing the location of the core, and for checking for damage of the honeycomb.

Holography Optical holographic interferometry is a technique used for measuring minute surface displacements. When load is applied to a bonded assembly, defects become apparent as local perturbations in the holographic interferogram. The technique is really only applicable to thin adherend structures. Acoustic holography uses pulsed echo ultrasound and focused transducers to enable a hologram of reflections from the bondline. With both techniques the equipment required is very expensive and the process time-consuming.

5.4 Safety

No discourse on the specification and use of adhesive materials would be complete without referring to safe handling precautions.

Safety aspects should partner the measures practised within a quality system, and they represent a vital part of the education and training of the personnel involved. The history of adhesive bonding has shown that accident and health problems associated with the technology are rare. Nevertheless, any material which is used without suitable precautions can become an unreasonable risk to health, safety, property, and the environment. The actual hazard presented by the use of materials depends not only on its potential hazard, but also on the conditions under which it is used. Concentration and duration of exposure to hazardous materials is generally the critical element in determining risk factors.

Legislation requires the adhesive supplier to label and classify products (as flammable, irritant, harmful, corrosive, or poisonous), as well as including standard risk phrases and safety procedures. Product information sheets must also be supplied in accordance with the Health and Safety at Work Act, detailing a large range of product data and safe handling precautions.

Many engineering adhesives contain no solvents and therefore problems such as solvent abuse or 'sniffing' are not present. However, solvents are sometimes used in considerable volume for surface degreasing processes, as well as for cleaning-up operations. Strict control must therefore be exercised over such procedures. Some surface pretreatment processes require the use of acids and other chemicals, for which a hazard is presented both in handling the liquids as well as with the proper disposal of the residues. The common mechanical abrasion techniques used for removing surface detritus and for preparing surfaces (such as grinding, blasting, and so on) also carry an obvious, albeit limited, risk.

On the whole, 'commonsense' precautions such as the use of skin and eye protection are sufficient for many applications. It should be remembered that the risks inherent in other surface joining processes such as brazing or welding can be very large, both from the high energy input required and from the resulting fumes. The use of adhesives can therefore improve significantly the safety in certain joining processes.

PART TWO

Applications

CHAPTER SIX

Applications in repair and strengthening

6.1 Concrete repair

The 1950s and 1960s saw an enormous growth in new construction and many faults are now becoming evident in some of the structures constructed during these years of peak activity. Today, repair and refurbishment is often preferred to demolition and reconstruction with the result that the maintenance and refurbishment market now represents approximately 50% of the construction industry workload of £37bn per annum. Concrete repair is no exception and currently represents a market of at least £500m annually. A variety of concrete repair methods have been developed many of which rely on adhesion between a repair material and the concrete substrate. The more important aspects of concrete repair will now be discussed under the headings of resin injection, patches, coatings and sealants and finally the implications of property mismatch.

Resin injection

The injection of resin-based materials into cracked concrete is a technique widely employed to either restore structural integrity or to seal the cracks for subsequent durability. However, a necessary preliminary to the consideration of resin injection is an assessment to establish the cause of cracking in the first place. Here it is important to distinguish between inactive cracks which no longer move and live cracks which may continue to move with changing loads or temperatures.

In order to assist in establishing the cause of cracking in concrete it is often helpful to distinguish between intrinsic cracking(1) and that caused by an externally applied stress. Intrinsic cracking can then be further subdivided by considering the time period over which the cracking is likely to occur:

(1) Plastic settlement and plastic shrinkage cracks – occur within a few hours of casting.
(2) Early thermal contraction cracking – occurs within a few days of casting.
(3) Cracking due to drying shrinkage – occurs within a few months of casting.
(4) Cracking due to reinforcement corrosion or alkali – aggregate reaction – may take many years to develop.

The most common forms of cracks due to structural loading are those transverse to the direction of the main reinforcement, caused by direct or flexural tension, and diagonal shear cracking.

The existence of a crack or cracks in reinforced concrete members does not necessarily imply that they need to be repaired. Indeed, most members are designed such that controlled cracking is expected under normal loading conditions. The need for repair thus depends on the crack width, the exposure conditions and the direction of the crack. Those cracks which run parallel to the direction of reinforcing steel are particularly important as they may expose long lengths of bar to possible corrosion. In many cases such cracks may be an early sign of reinforcement corrosion already in progress and simple sealing will only provide a very short term remedy(2).

Cracks which have been identified as inactive can be sealed by injection with an appropriately formulated low viscosity resin. If a structural bond is required across the crack faces then epoxy resins are most suitable. For simple sealing against the ingress of moisture, alternatives include polyester and acrylic resins although these are likely to undergo greater initial contraction than epoxies. Crack widths down to 0.1 mm can be successfully filled using resin injection provided the crack surfaces are clean and sound. Those factors affecting the choice of resin will include:

(1) the required pot-life at the application temperature;
(2) the crack width and area which has to be filled;
(3) any requirement to displace water in the cracks;
(4) the rate at which strength must be developed at the anticipated ambient temperature and humidity prior to application of load;
(5) whether or not a structural bond to the crack faces is required.
(6) the method of injection.

Although it is possible to use straightforward gravity to fill some cracks, usually some form of injection under pressure is more satisfactory. Resin is injected at the lowest point so that air and any water it displaces flow upwards. The external faces of the crack are temporarily sealed between injection points to prevent the resin draining out before it has hardened. Alternatively, thixotropic formulations can be employed in situations where it is difficult to guarantee effective sealing along all external faces. Injection techniques vary from simple hand held cartridges to sophisticated machines incorporating metering and mixing devices. The selection of an appropriate pressure for injection is usually based on the need to ensure complete filling of the crack without being so excessive as to cause further crack opening. For the impregnation of highly cracked surfaces vacuum techniques can be employed to remove most of the air in the cracks whilst resin is introduced under atmospheric pressure.

For cracks which have been identified as live the options available are structural repair to prevent further movement followed by resin injection, or widening of the crack surface and treatment as a movement joint. Surface widening of cracks is best performed with a crack router or high pressure water jet. The groove so formed is then sealed with a low modulus material whose strain capacity is compatible with that developed across the joint during subsequent crack movement(3). Use of a rigid material in live cracks will result in crack redevelopment in the adjacent concrete. If the resulting jagged nature of a crack repaired in this way is visually unacceptable a more expensive alternative is to repair the crack using injection techniques and to construct a straight movement joint as close as possible to the original crack line.

Patches

Spalling of the surface layers of concrete in reinforced members may occur as a result of reinforcement corrosion (Fig. 6.1), the effects of fire or due to impact. Where the spalling is local and the cover is more than 25–30 mm, conventional sand–cement mortars based on high quality sands and applied using good rendering practice may be used. However, great care in surface preparation is necessary to ensure adequate adhesion.

For local spalling where the cover is between 12 and 25–30 mm, polymer modified cementitious materials are more appropriate. The polymers are usually supplied as very fine polymer particles dispersed

Fig. 6.1. Spalling of concrete in reinforced members.

in water to form a stable white milky emulsion or latex. In some cases, however, the polymer can be supplied in fine particulate form. The polymer modification of the mortar can be beneficial for the following reasons:

(1)   it acts as a water-reducing agent;
(2)   it improves bond to the concrete substrate;
(3)   it improves flexural and tensile strength;
(4)   it reduces permeability to water and carbon dioxide;
(5)   it acts to some extent as an integral curing aid.

Polymer types include (refer to Chapter 2 for further details):

Polyvinyl acetate (PVA) – not recommended in external applications or under wet service conditions.
Polyvinyl dichloride (PVDC) – not recommended for use with reinforced concrete because of the danger of chloride release.
Styrene butadiene rubber (SBR) – suitable for most circumstances.
Acrylics and modified acrylics – suitable for most circumstances.

A typical repair mortar composition might comprise:

200

Concrete repair

	Parts by weight
Ordinary Portland cement	100
Clean sand to BS 882 (4) (5 mm particle size and below)	250–300
Polymer latex (Approximately 50% solids content)	20–30
Water	The minimum to give the required consistency

To improve adhesion to the prepared concrete surface, bond coats comprising a polymer latex and cement mixed to a slurry may be applied. However, the mortar must then be trowelled in place whilst the bond coat is still tacky or adhesion problems at the bond coat/mortar interface may develop with time. Epoxy resin bond coats are often used on the reinforcement following cleaning using gritblasting or water jetting. The epoxy may be lightly dashed with sharp sand to improve the mechanical key with the subsequently applied mortar. Some resin bonding aids are formulated to receive the cementitious repair before the resin cures. Fig. 6.2 illustrates schematically a typical preparation procedure and the process of patch repair of spalled concrete.

For the repair of shallow spalls or where the cover to the steel is less than 12 mm repair mortars based on reactive thermosetting resin binders may be more appropriate. Their advantages include:

(1) Development of high mechanical strength within 24–48 hours.
(2) Encapsulation of the reinforcement steel within a low permeability barrier which has excellent adhesion to both the reinforcement and concrete substrate.
(3) The possibility of formulation with lightweight fillers to produce low density thixotropic mortars which assist in repairs on vertical and overhead surfaces.

Resin types include (refer to Chapter 2 for further details):

Epoxies – formulated with specially graded dry fillers. Correct proportioning and mixing are vital to achieve optimum properties. Because of the exothermic reaction thermal contraction on cooling may cause interfacial stresses to be built up. However, with epoxies the maximum heat evolution occurs whilst the resin is in a fluid state. With careful formulation and correct mixing and application the heat output can be controlled and shrinkage becomes negligible.

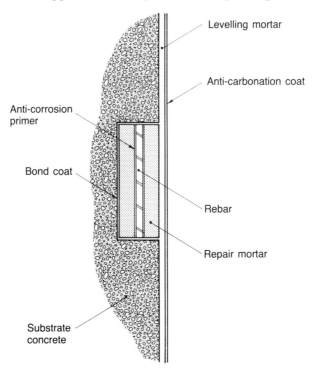

Anti-corrosion primer

Bond coat

Substrate concrete

Levelling mortar

Anti-carbonation coat

Rebar

Repair mortar

Fig. 6.2. The patch repair process.

As has been mentioned in Chapter 2 the rate of curing of epoxy systems is temperature dependent and thus it is important to select a formulation appropriate to the ambient temperature during application. At 20 °C most epoxies develop 80% of their ultimate properties within 24–48 hours.

Unsaturated polyesters – proportioning and mixing is less critical than with epoxies. Again, curing is an exothermic reaction but the maximum heat evolution occurs after the resin has set. This can result in significant thermal contraction inducing high stresses and maybe debonding at the interface with the substrate concrete. In addition the volume of the hardened polymer is less than that of the freshly mixed adhesive. For these reasons polyester mortars are only applicable for small volume repairs but they have the advantage of being able to cure at lower temperatures than epoxies and thus may be used overnight for load application the next day.

Acrylics (methyl methacrylates) – tend to be formulated such that they exhibit less shrinkage than conventional polyesters. However, they have low flash points and are therefore highly inflammable in their pure resin state.

Many resin-based materials have a relatively low glass transition temperature and will lose a significant proportion of their strength and stiffness at this temperature. For the repair of fire damage internally within a building it is recommended that resin repairs are only used when either:

(1) Performance data demonstrates adequate fire resistance such that the material retains its structural properties under future envisaged fire conditions; or
(2) The material is adequately protected from fire by other materials and retains its structural properties at the expected fire temperature at the relevant depth within the section; or
(3) Loss of strength of other properties will not cause an unacceptable loss of structural section or fire resistance.

With resin-based repair mortars the reinforcement bars should be gritblasted and the system applied within four hours. Otherwise a non zinc-rich holding primer may be considered. Rust converters should be treated with caution since some containing phosphoric acid can inhibit cure of the subsequently applied resin. The top coat of any compatible primer or bond coat applied to the substrate concrete or steel bars should still be tacky when the repair mortar is applied.

Coatings and sealants

In general a coating in the context of concrete repair can be considered as a fluid applied to the surface which forms a continuous film. In so doing the coating must be capable of adhering effectively to the concrete substrate. The main purposes of using coatings on concrete, apart from reasons of appearance, are to control water absorption and the passage of vapour or to act as a barrier to protect it from aggressive chemicals.

The use of coatings on concrete has shown a marked increase in recent years in order to resist carbonation or to control the ingress of chlorides. These factors have been responsible for much of the reinforcement corrosion and surface spalling of concrete which is now evident.

Carbonation of concrete is the result of carbon dioxide gas diffusing into the concrete pores. Carbonation rates tend to be highest when the concrete is relatively dry since the pores contain little water to prevent entry of the gas but just sufficient to allow it to dissolve. Carbonic acid is formed which reacts with the free lime in concrete to form calcium carbonate and leads to a gradual fall in alkalinity from the surface inwards. Once the carbonation front reaches the steel reinforcement depassivation occurs and, in the presence of water and oxygen, corrosion can proceed.

Anti-carbonation coatings are thus designed to resist the diffusion of carbon dioxide and oxygen into the concrete. In addition they should allow the free passage of water vapour so that vapour pressure does not build up behind the film and cause loss of adhesion. The resistance of a coating to the diffusion of carbon dioxide is related to its diffusion resistance coefficient μ_{co_2} a dimensionless parameter indicating how many more times resistant a coating is than static air. The product of μ_{co_2} and the dry film thickness, s, gives an equivalent air layer thickness which it is normally accepted(5) should exceed 50 m for good anti-carbonation properties. Similarly the product of μ_{H_2O} and s should be less than about 4 m to avoid build up of water vapour pressure behind the coating. Examples of anti-carbonation coatings include acrylic emulsions which cure by drying, solvented chlorinated rubber systems and polyurethane resins which undergo a chemical cure by use of a catalyser. All may be supplied in pigmented form. However, if the concrete contains active cracks then the coating must be sufficiently thick and flexible to bridge the cracks and prevent carbonation near them. Examples include mastic asphalt on bridge decks and some high-build polyurethane formulations.

Chlorides penetrate concrete from the outside only when they are in solution in water. Even quite small amounts of chloride, e.g. 0.4% by weight of cement, can disrupt the oxide layer on the surface of steel reinforcement bars which normally inhibits corrosion. Thus coatings which resist water penetration will also serve to resist the ingress of chlorides but at the same time the concrete must be allowed to 'breathe'. A number of coatings or sealers including epoxies, methacrylates, urethanes and chlorinated rubber are currently the subject of test programmes to assess resistance to chloride ingress. In addition water repellant pore liners such as silicones, siloxanes and silanes, which work by altering the surface tension in the pores, and pore blocking materials such as silicates and crystal growth materials are under study.

At present it is not possible to give advice based on broad generic type since performance tends to be product-specific.

In certain locations concrete structures may be subjected to particularly aggressive acidic environments. Examples include acid rain from release of sulphur dioxide into the atmosphere and culverts conveying water containing dissolved organic acids. In such circumstances coatings may need a fairly high degree of chemical resistance in addition to anti-carbonation properties. Epoxy paints or high-build epoxy/pitch coatings typify the type of material used to afford protection.

The efficiency of any coating depends on adequate adhesion to the substrate concrete and the normal rules for achieving this will apply. The surface must be clean of any oil or grease such as that used in mould release agents. Water jetting or steam cleaning may be necessary on old concrete, with the surface allowed to become reasonably dry prior to painting or spraying on liquid coatings. In many cases priming systems are recommended to promote adhesion of the coating system.

Property mismatch

For the surface screeds and coatings applied for purely cosmetic purposes the repair will have little influence on the overall structural performance of the repaired member. However, with maintenance programmes now being formulated to provide the longest possible life before major renovation becomes necessary it is likely that patch repairs when they are carried out will be of a scale where structural integrity becomes of importance. Work in the USA(6) using cementitious repair mortars has shown the importance of taking the patch well behind the reinforcement steel to assist in ensuring the overall integrity of a reinforced concrete beam. Current UK guidelines(2) confirm that, for the most durable patch repairs, concrete should be cut away right around the affected reinforcement.

Possible concerns arise over the potential property mismatch between the repair material and the substrate concrete. Short-term problems may arise because the repair material contracts on curing relative to the surrounding concrete. With resin based mortars the

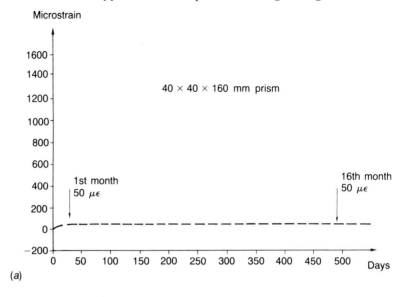

(a)

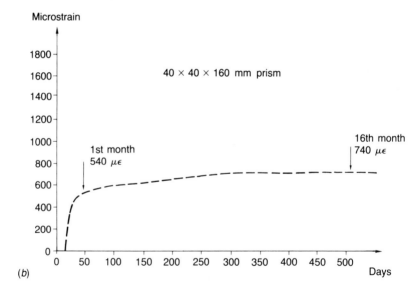

(b)

Fig. 6.3. Free shrinkage strains in repair mortars. (*a*) Epoxy mortar. (*b*) Styrene Butadiene Rubber modified mortar. (*c*) Magnesium phosphate modified mortar. (*d*) Ordinary Portland cement mortar.

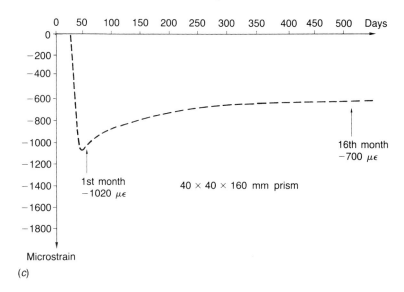

(c)

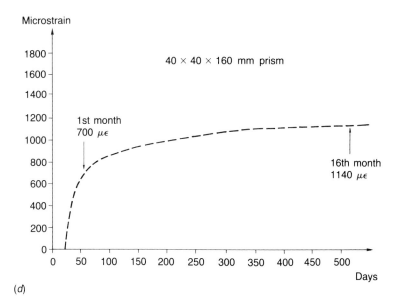

(d)

Table 6.1. *Mechanical properties of selected systems*

A. Epoxy mortar
B. Polyester mortar
C Acrylic (resinous) mortar
D. SBR latex modified mortar
E. Vinyl acetate (cementitious) mortar

F. Magnesium phosphate mortar
G. Sand/cement mortar
H. High alumina cement mortar
I. Flowing concrete

Repair system

Mechanical property	A	B	C	D	E	F	G	H	I
Comp. strength N/mm²	69.0	97.3	71.0	79.5	57.8	74.3	83.0	77.1	106.3
Tensile strength N/mm²	13.7	14.7	10.1	6.5	2.6	4.2	3.1	4.1	3.0
Flexural strength N/mm² 4-point bending	21.9	27.5	14.3	12.5	7.9	8.6	5.6	6.2	6.5
Flexural strength N/mm² 3-point bending	27.8	36.0	18.2	14.4	7.7	10.0	10.1	8.5	7.8
Comp. modulus kN/mm²	18.8	24.8	17.9	35.4	18.6	49.4	26.7	36.0	40.7
Tensile modulus kN/mm²	23.7	25.4	13.9	43.1	31.9	59.8	38.6	28.0	46.4
Flexural modulus kN/mm² 4-point bending	13.9	20.7	10.5	24.8	25.1	37.3	22.6	27.8	31.3
Flexural modulus kN/mm² 3-point bending	15.9	18.6	7.5	21.4	13.2	13.5	14.2	16.3	12.4
Coef. of thermal Expan./contrac. × 10⁻⁶ (°C⁻¹)	16.4	19.8	19.1	14.3 (−60 °C TO +20 °C)	12.0	11.9	5.8	10.4	11.5
	36.8	30.1	19.1	10.9 (+20 °C TO +60 °C)	12.0	11.9	9.4	10.4	11.5

prime cause is the cooling which occurs following the exothermic reaction whereas with water-based cementious formulations drying shrinkage may occur. Fig. 6.3 shows the free shrinkage strain measured on 40 mm square by 160 mm long prisms of four different generic forms of concrete repair mortars, measurements starting when the materials had attained an age of 24 hours. Of particular note is the relatively low shrinkage on cure observed with the epoxy mortar as compared with some cementitious systems. Also of interest is the expansion observed with a magnesium phosphate system. This is deliberately induced by the incorporation of an expansive admixture(7) in the formulation. The volume decrease which occurs with polyester based systems on hardening has been referred to earlier in this chapter and must be considered in addition to the curing shrinkage. The manifestation of curing contraction is either initial tensile strains induced in the repair or cracking at the repair/substrate interface, both of which may reduce longer term structural capacity.

During service, incompatibilities in the form of differing elastic moduli and differential thermal movements between repair and substrate may create problems. The strengths, elastic moduli and coefficients of thermal expansion for nine repair systems are summarised in Table 6.1(8). The results show that all systems provide adequate compressive strength for the vast majority of applications, the values generally being well in excess of those expected in the substrate concrete. The resin-based materials have greater tensile and flexural strengths than the polymer modified cementitious systems which in turn have greater values than unmodified ordinary Portland cement materials and also of the substrate concrete. Inspection of the modulus of elasticity results, however, shows a reverse trend, the resin based systems generally having lower moduli than cementitious systems. Indeed, the moduli of some cementitious systems are in excess of those expected in the substrate concrete.

To illustrate the effect of modulus mismatch on structural performance consider the symmetrically repaired reinforced concrete prism shown in Fig. 6.4 and loaded axially in tension. The concrete has a modulus of elasticity in tension of 25 kN/mm^2. For material C ($E_t = 14$ kN/mm^2) the elastic stress induced in the concrete at mid-height of the prism will be 2.5 times that in the repair material at the same position. Conversely for material D ($E_t = 43$ kN/mm^2) the elastic stress carried by the repair material is now 1.5 times that

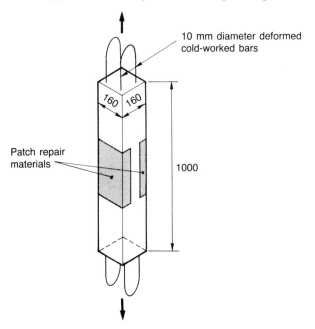

10 mm diameter deformed
cold-worked bars

160 160

Patch repair
materials

1000

Fig. 6.4. Axially loaded repaired prism.

carried by the concrete. This enhanced stress must be transmitted across the interfaces at the end of the repaired zone thus increasing the demands placed upon the repair/substrate adhesion. In addition, the elastic property mismatch at the end interfaces of the repaired zone can induce zones of stress concentration. For a low modulus repair material these occur on the surface of the stiffer concrete immediately adjacent to, but not at, the interface (Fig. 6.5). If the peak tensile stresses exceed the tensile capacity of the concrete then cracking will occur and the possibility of subsequent reinforcement corrosion being initiated outside the repaired area must be considered. This point highlights the need for careful consideration of where patch repairs are curtailed with respect to the overall stress state in a structural member.

It can also be seen from Table 6.1 that the coefficients of thermal expansion/contraction of resin-based repair systems may be two or three times that of the substrate concrete. If the repair is carried out at, say, 10 °C then a rise in ambient temperature to 25 °C will potentially induce a compressive strain in the repair material of the

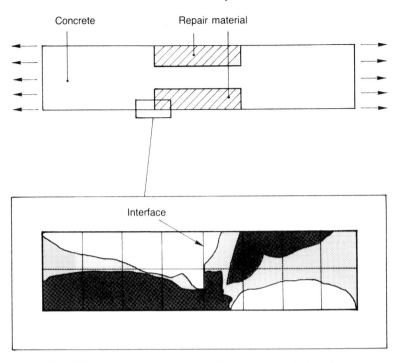

Fig. 6.5. Stress concentrations adjacent to repair interface.

order of 400 microstrain. Conversely, tensile strains will be induced in the repair material during winter months if the repair has been carried out during the summer.

A further major consideration in the repair of structural concrete elements is the decision as to whether to relieve the element of load whilst the repair process is being performed. If the application of live or imposed load is prohibited during the process of concrete removal but dead load is not temporarily resisted by propping then there will be a redistribution of dead load stresses within the residual section. Further, the patch repair material will not subsequently contribute to the support of dead load. Alternatively, if both dead and live loads are supported by temporary props whilst the repair is carried out then the reapplication of dead load will result in subsequent creep of the repair material under the sustained stress, whilst any cyclic live loads will induce fatigue loading on the repaired section. A third possibility, for example in the case of columns

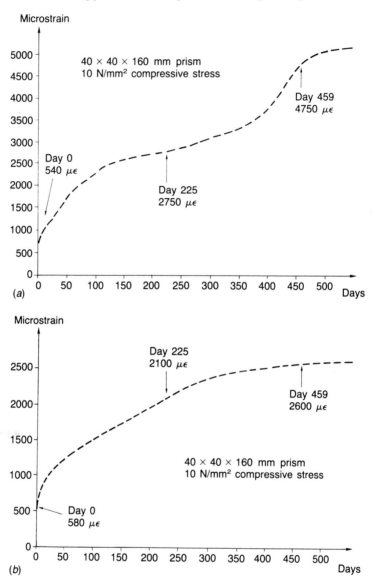

Fig. 6.6. Compressive creep strains in repair mortars. (*a*) Polymethyl methacrylate mortar. (*b*) Polyvinyl acetate modified mortar. (*c*) Magnesium phosphate modified mortar. (*d*) Flowing concrete.

212

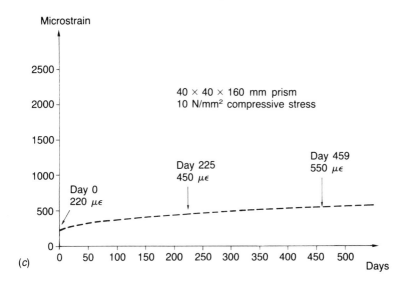

(c)

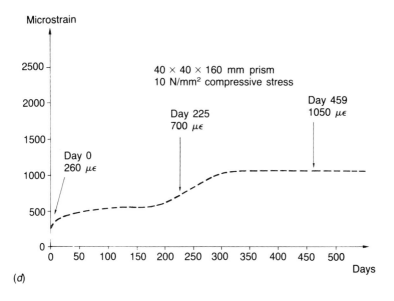

(d)

supporting a bridge deck, is where temporary trestles are erected to support the dead load of the deck but deflections induced by live loads apply a cyclic load to the columns during the repair operation.

The effect of a sustained compressive stress of 10 N/mm² on the creep strains induced by 160 mm long by 40 mm square prisms of four different generic forms of repair material is shown in Fig. 6.6. The range of ultimate creep strains varies from over 4500 microstrain with one resin-based system down to less than 400 microstrain with some modified cementitious systems. These strains include the effect of any curing shrinkage/expansion which may also be occurring during the period under load.

Consider, a simple scenario involving the 600 mm circular reinforced concrete column shown in Fig. 6.7. Corrosion of reinforcement necessitates the removal of surface concrete to a depth of 25 mm behind the reinforcement. The column is subject to design axial loads of 2000 kN dead and 1200 kN imposed, both of which are fully relieved by propping whilst the patch repair is effected in order to prevent overstressing the remaining concrete core. A repair with the relatively low compression modulus modified cementitious mortar 'E' (Table 6.1) causes a 24% increase in concrete stress to 12.5 N/mm² whereas with the high modulus flowing concrete 'I' there is a 13% decrease in stress. The corresponding stresses in the repair materials are 7.7 and 11.5 N/mm², respectively.

Taking creep into account, a repair with material 'E' loaded at an age of 1 month will cause the dead load stresses in the concrete

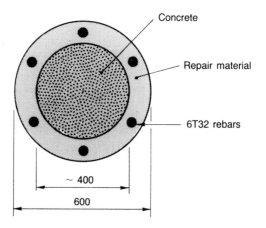

Fig. 6.7. Repair of surface concrete in circular column.

214

to increase from 7.9 to 10.8 N/mm^2 with time. The latter value increases to 15.5 N/mm^2 when live load is applied which may cause some concern in relation to the ultimate strength of the concrete. Meanwhile, the dead load stresses in the repair have reduced from 4.8 to 2.2 N/mm^2.

It is thus apparent that care must be taken when selecting materials for the deep patch repair of reinforced concrete members to ensure that the material is structurally compatible with the substrate.

6.2   Concrete strengthening

It is sometimes necessary to enhance the serviceability performance or ultimate load-carrying capacity of certain structural members in existing concrete structures. This might include the control of cracking in reinforced concrete beams and slabs under the action of superimposed loads, or a need to increase the reinforcement due to design deficiencies or construction errors. Increases in ultimate load-carrying capacity may be required due to amendments in the design codes or changes in use of the structure either for a one-off loading or for continuous use.

Conventional techniques for dealing with such situations involve the construction of additional supports, prestressing or the enlargement of structural members. In the latter case this may involve the bonding of new concrete to old. An alternative solution is to use mild steel plates bonded by an adhesive to the external surface of the concrete member in question. These two techniques will now be discussed in turn.

Concrete overlays

Unless a portion of the original concrete is removed and replaced with a higher strength material, then a concrete overlay will usually result in an increase in member self weight. Applications at ground level are therefore more common than in suspended beams or slabs, although the use of polymer modified concrete overlays has found some favour in the refurbishment of bridge decks in North America(9).

Adhesives are often employed to achieve the necessary longitudinal shear connection for full composite action between the fresh and the older hardened concrete. This is because the inherent bond between fresh and hardened concrete cannot be relied on for a truly structural connection. The uncertainty of the purely cementitious bond has been related to the early occurrence of shrinkage stresses and the relatively slow rate of bond strength development. There are two prerequisites of a strong bond. The first is the careful preparation of the old concrete surface; removal of laitance and surface contaminants by grinding or gritblasting is imperative. The second is the selection of an appropriate adhesive.

Epoxies are preferable to other resin systems in that they can be formulated to bond to moist hardened concrete and yet be able to cure within the wet environment of the fresh concrete. Polyesters do not generally bond reliably under wet conditions. The system must also be selected with due regard to the temperature of application. Resins claimed to have good bonding performance at 20 °C may give rather poor bond strengths at the lower cure temperatures prevalent on UK construction sites during much of the year.

There are several records of tests carried out to demonstrate that epoxy adhesives can achieve good bond strengths with failures occurring in the substrate concrete rather than within the glue-line(10, 11). The main danger to the long-term integrity of the bond is the presence of water and thus water penetration to the glue-line should be prevented as far as possible.

In recent years the use of polymer latex slurries as bonding coats between new concrete and old has been the subject of continued debate. It is now universally recognised that PVA systems are unsuitable in damp conditions. SBR bonding agents have been utilised since the late 1960s on many UK construction sites whereas acrylic based systems are a more recent addition. The arguments have largely centred around the method of test and the effect of multiple coats. The slant shear test is recommended in BS 6319 (12) but it must be recognised that in this test the bond line is subject to compressive as well as shear stresses. Further, the result is sensitive to the method of preparation of the hardened concrete surface (see Chapter 4).

With SBR slurries the use of more than one coat can reduce bond strengths assessed using the slant shear test(13). It is now accepted that these materials form a skin if allowed to dry which can impair adhesion of subsequent layers.

Under normal conditions on repair sites it seems likely that a more reliable bond can be achieved by using a bond coat rather than by relying on any natural bond between fresh and hardened Portland cement concretes(2). However, bonding agents for use on site must be tolerant of site conditions, be reliable and cannot be selected solely on the basis of slant shear results. The bonding performance of carefully applied SBR and acrylic latex/cement slurry coats appears to be similar to that of epoxy resin bonding aids(14). However, other factors such as open time, barrier coat effects and cost may affect the final choice.

Externally bonded reinforcement

The advantages of externally bonded reinforcement over other methods of strengthening concrete structures include the ability to strengthen part of a structure whilst it is still in use, minimum effect on headroom, low cost and ease of maintenance (see Fig. 6.8, for example). The method has been in use for over 20 years, mainly to enhance flexural capacity, and has been found to produce effective and economical solutions to particular problems.

Historical development. The technique was introduced in France in the late 1960s where the first reported application was a major

Fig. 6.8. Strengthening of a motorway structure whilst still in use.

217

bridge strengthening scheme on the Autoroute du Sud. It was also used in several other countries during the early 1970s including Switzerland, South Africa and Japan. For example, up to 1975 well over 200 bridges in Japan had been strengthened to accommodate the large increases in loading that had occurred due to increases in heavy goods traffic since the bridges were designed.

In 1975 the first application to a bridge in the UK occurred when 2 pairs of motorway bridges on the M5 at Quinton were strengthened after cracks which were described as 'little more than shrinkage cracks' were observed during a routine bridge inspection. A detailed appraisal of the design showed that the cracking occurred as a result of stopping off too many reinforcing bars in a zone centred at the quarter points of the main span. Loading tests, carried out before and after strengthening, demonstrated the effectiveness of the technique in providing increased flexural stiffness and in reducing crack opening under load(15).

Two bridges on the M20 in Kent were strengthened shortly after construction in 1977. Cracks were discovered in one of the side spans which were found to have insufficient longitudinal steel. Steel plates were subsequently bonded to the side span soffits and the top surface of the deck over the supports (Fig. 6.9).

The floor beams of a North London building were strengthened in 1978 to allow for increased floor loadings following a change in use. The factor of safety provided by the existing structure, in the

Fig. 6.9. Swanley bridge during plating operations.

event of fire rendering the adhesive bond ineffective, was calculated to be 1.1 and this was deemed to be satisfactory.

Meanwhile the technique was gaining ground overseas. The capacities of solid slab floors in two telephone exchanges in Zurich were raised from 2 to 7 kN/m² to accommodate new switching equipment. Here, constraints which led to adoption of the technique in preference to other strengthening methods were that there had to be no substantial reduction in room height and the need for minimum delay in reconstruction. Other examples include the strengthening of the roof structure at the Central Railway Station in Warsaw and as part of a structural repair process following blast damage within an apartment of a twenty-six storey building in Brussels. The only known North American application of the technique involves a modern University building in Canada. The floor slabs of a computer room were reinforced in bending to accommodate an increase in loading. Supporting beams were also strengthened using steel strips bonded to the vertical faces to comply with the requirements for shear in the latest code of practice.

A precast, prestressed hollow-box beam skew bridge on the M1 in Yorkshire was required to support a vehicle carrying a generator to Cumbria. The total load of the lorry plus generator was 460 tons. Strengthening the bridge using an additional overlay slab of reinforced concrete was considered but this would have resulted in considerable increases in dead weight and additional road works. Another possible solution considered was to strengthen using longitudinal plates which would have increased the bending strength, but would have done little to enhance the insufficient shear capacity of the bridge. The solution actually adopted was to bond plates to the soffit of the deck in a direction parallel to the abutments which was 45° to the longitudinal axis of the hollow box beams. The strengthening was successful as only small deflections of the deck were observed during the passage of the abnormal load over the bridge.

Since 1983 there have been, on average, one or two applications of bonded plating to buildings per year in the UK. Following a fire in a single-storey school building in Glasgow, prestressed concrete roof beams were redesigned and strengthened using flexural plates, assuming zero residual prestress. Also in Glasgow, concrete roof beams in a distillery were strengthened in flexure and 'nominally' in shear following removal of supporting internal walls during reconstruction (Fig. 6.10). The cracking which developed in a West

219

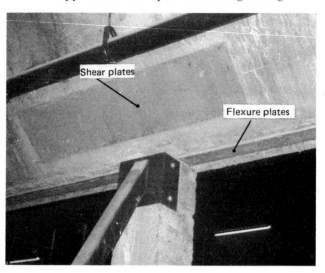

Fig. 6.10. Strengthened beam within a Glasgow distillery.

Midlands police car park soon after construction, as a result of a deficiency in conventional reinforcement, has been well publicised. Here again, additional bonded plate reinforcement was installed in order to control further cracking under live load. An 11 m span, 2 m deep reinforced concrete beam at a shop in Bootle was strengthened to provide a 10% increase in ultimate moment capacity. In so doing it was calculated that the mean horizontal shear stress resisted by the adhesive was 9.84 N/mm².

In 1985 cracking was noticed in the floor slabs of a multi-storey office building in Leeds. The cracks were adjacent to the external columns and the central lift well and design checks indicated a deficiency in both shear capacity and top flexural reinforcement. A combination of soffit supporting brackets and steel plates bonded to the top surface adjacent to supports was adopted to restore capacity and control cracking (Fig. 6.11). Subsequent load tests revealed that the steel plates were attracting tensile stresses up to 40 N/mm² at 1.35 times design load.

A recent example of a bridge strengthening scheme in this country is the upgrading of a pedestrian bridge at a service station on the M2 in Kent. After design checks revealed that there was insufficient steel in the deck and a bridge inspection showed that there were transverse cracks in the soffit, longitudinal plates were applied to

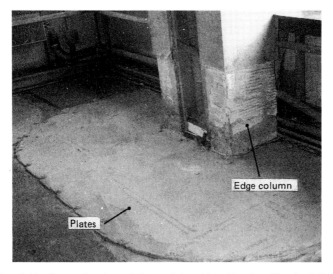

Fig. 6.11. Strengthening of floor slab within Leeds office building.

the soffit at each end of the bridge to provide the extra strength required. In 1987 an access ramp and bridge were strengthened at Felixstowe docks to accommodate the increased axle loading now required by EEC regulations.

These examples illustrate the range and type of applications which are likely to benefit from steel plate bonding. All have been carried out successfully and the required extra strength has been achieved.

Structural design. For applications involving the addition of steel plate reinforcement to enhance the flexural capacity of an existing reinforced concrete structure, design calculations may be based upon normal reinforced concrete theory. One of the advantages of the technique over other forms of strengthening is the ability to carry out the work while the structure is still, at least partially, in use. Thus, any temporary propping of the structure to relieve dead load stresses is unlikely to be attractive. The calculation process under design working loads is best carried out in stages:

(1) Calculation of existing stresses based upon the dead and permanent loads acting on the original section.
(2) Calculation of additional stresses based upon the imposed or live loads acting on the strengthened section.

(3) The stresses are added algebraically and a check made to ensure that those in the concrete and both sets of reinforcement are acceptable.

(4) Calculation of the interfacial shear stress in any adhesive layer and a check made to ensure that this is acceptable.

For the purposes of these calculations steel areas may be transformed into equivalent concrete areas by using a modular ratio adjusted for permanent loads where appropriate. It should be noted that, although the effect of plating is to change the position of the neutral axis and hence to increase the area of concrete resisting flexural compression, the need to limit the stresses within the existing sections may govern the design. A final check should be made to ensure that the moment of resistance of the section under ultimate conditions is adequate and that a ductile failure mode would occur.

A well prepared adhesive joint using a cold cure epoxy resin will have an intrinsic shear bond strength of between 10 and 20 N/mm^2 depending on the adhesive. This is well above the shear strength of the concrete to which it is bonded which will probably be not more than about 4 N/mm^2. It is important that under ultimate conditions the local shear stresses in the adhesive do not anywhere exceed the shear strength of the concrete. For steel-to-steel joints, such as for multilayer plates or cover plates at butt-joints, a higher value of bond shear strength may be taken. Typical values of allowable mean shear stress in the adhesive layer for design under service loading are 1.2 N/mm^2 for a steel-to-concrete joint and 3.0 N/mm^2 for a steel-to-steel joint.

The plate thickness must be selected to allow reasonable flexibility for conforming with concrete surface irregularities. However, 3 mm is likely to be the minimum practical thickness to avoid distortion during gritblasting. A wider thinner plate gives better results because of reduced shear and normal stresses resulting from the bigger contact area but if the plate is too thin there is a risk of out of plane transverse bending which may cause the edges of the plate to lift from the surface. The most satisfactory results may be expected from plates having a width/thickness (b/t) ratio of about 60. This will ensure failure by yielding of the plate at ultimate load with little or no risk of sudden springing off of the bonded plate, while at the same time maintaining a significant stiffening effect and increase in failure load. Ratios of less than 40 should be avoided, particularly in continuous beams at regions of hogging bending moment where shear and bond stresses in the concrete are likely to be relatively high.

Wherever possible plates should extend over the complete length of the region in tension but in simply supported spans requiring flexural strengthening this is unlikely to be practical. In such cases the plates should extend over at least 80% of the span. Elsewhere, a minimum transmission length of 200 mm is a reasonable assumption. A contribution to plate anchorage will also arise if bolts have been used to support the weight of the plate during cure of the adhesive. If the above recommendations on plate geometry have been followed, the use of bolts is not strictly necessary but their use, particularly at plate ends, is a wise precaution against unexpected defects or poor workmanship. The contribution of bolting to plate anchorage is difficult to assess without a full finite element analysis since the distribution of both shear and normal stresses varies rapidly and will depend on the particular configuration of the plate relative to the structure. Typical stress distributions within the adhesive layer are illustrated in Fig. 6.12. It is generally safer to design an adequate anchorage based on adhesive bonding alone and to provide secondary anti-peel bolts on an empirical basis. It is recommended that such bolts be designed to resist a shear stress of three times the mean value over the effective anchorage area. The effective anchorage area is obtained from the product of the effective anchorage length (l_a) and plate width (b). The effective anchorage length increases with decreasing b/t ratio and may be obtained from Fig. 6.13. Such bolts will also be capable of resisting any normal stresses at the plate ends.

More than one layer of plates may be needed to transmit the required load and yet meet the width/thickness ratio requirements for individual plates. Alternatively, it may be necessary to form a butt-joint between plate ends in which case the load must be transferred through a bonded cover plate. Such joints should be kept to a minimum since they result in a change of section stiffness. They should also be avoided at locations of high deformation, e.g. structural plastic hinges, and in the region of concrete construction joints. Cover plates should have the same dimensions as the main plates and overlay lengths must be sufficient to minimise the force which the concrete is called upon to transmit. Research(16) suggests that the total overlap length for plates having the recommended b/t ratio of 60 should be at least 400 mm.

An environment which combines high moisture levels and de-

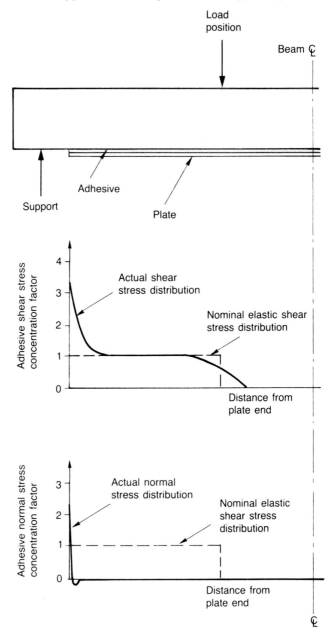

Fig. 6.12. Stress distributions within adhesive layer.

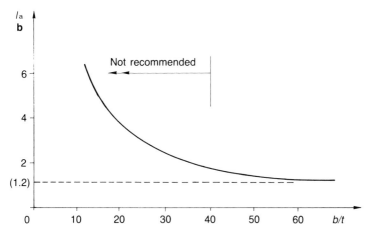

Fig. 6.13. Effective anchorage length at plate ends.

icing salts with elevated temperatures is likely to be particularly deleterious to the long-term durability of the adhesive bond. Research(17) suggests that many of the two component, cold-cure epoxy resin adhesives which have been used for bridge strengthening purposes are likely to suffer a significant reduction in elastic properties and bond strength above the temperature range 40–50 °C or after long periods of immersion in water. However, mechanical properties are largely recoverable if the rise in temperature has been caused by an isolated extreme in ambient conditions.

Requirements of the adhesive. A full compliance spectrum for steel/concrete bonding has been published by the authors(18) and is reproduced as an Appendix at the end of the book. The purpose of the adhesive is to produce a continuous bond between steel and concrete to ensure that full composite action is developed by the transfer of shear stress across the thickness of the adhesive layer. Experience has shown that the best chance of success is likely to be achieved by using cold-cure epoxy based adhesives which have been specially developed for use in the construction industry. Provided that the surfaces have been prepared properly, these bond well to both steel and concrete and do not suffer shrinkage and cracking problems such as may occur with other systems like polyesters. For these purposes a cold-cure adhesive is defined as one which is capable of curing to the required strength between the temperatures of 10 °C and 30 °C. The resin component will normally be based upon diglycidyl ether of 'bisphenol A' or 'bisphenol F', or a blend of the two. The hardener, or curing agent, will normally be from the polyamine group, since these tend to produce adhesives with better resistance to moisture than do the polyamides and are less likely to give concern over creep performance under sustained load than with polysulphides. Other additives such as diluents, flexibilisers, plasticisers, toughening agents and inert fillers may also be incorporated into the formulation to improve the application or performance characteristics of the adhesive. In the case of inert fillers these may alternatively be supplied as a third component for inclusion at the time of mixing. It is important that the filler be of a non-conductive material, be highly moisture resistant, be capable of withstanding temperatures well in excess of maximum service temperature and typically shall have a maximum particle size of 0.1 mm. This latter requirement is to minimise the possibility of moisture penetration around the surface of the particles. The toxicity of the chemicals used in the adhesive and any associated primer for use on the steel surface must be low enough to enable safe use on a construction site.

To ensure thorough mixing it is helpful if the resin and hardener are of dissimilar colour. The adhesive should also mix to a smooth paste-like consistency suitable for spreading on both vertical and horizontal surfaces of either concrete or treated steel, in layers from 1 to 10 mm thick to allow for concrete surface irregularities. The pot-life of the mixed adhesive, which determines the time after mixing within which it must be used before it starts to harden, generally needs to be at least 40 minutes at 20 °C. A similar joint open time, which represents the time limit during which the joint has to be closed after the adhesive has been applied to the surfaces to be joined, is also necessary. If the joint is closed after this time the strength of the bond may be dramatically reduced because the exposed surface of the hardener reacts with moisture and carbon dioxide in the atmosphere in a way which impairs adhesion. This effect can be minimised by roughening the surface layer of the wet adhesive just before closing the joint. Since both pot-life and open time are temperature and humidity dependent it is evident that the sequence of operations must be planned carefully to ensure that the adhesive can be applied and the joint completed within the allowable times. For applications involving repair or strengthening the adhesive usually needs to be capable of curing sufficiently to give the required mechanical properties within a period of 3 days at 20 °C in relative humidities up to 95%.

The necessary mechanical properties can be sub-divided into those required of the hardened adhesive itself and those representing bond efficiency with appropriate adherends. In the former case, limiting properties for moisture and temperature resistance together with minimum values of flexural modulus and shear strength are suggested in Table 6.2(18). The limit on water uptake by the adhesive is to assist durability of the steel/adhesive or concrete/adhesive interface even if moisture uptake is not deleterious to the adhesive itself. The lower limit for heat deflection temperature is to ensure that maximum likely ambient temperatures in the UK do not affect the efficiency of the bond. The minimum value for flexural modulus is to guard against problems due to creep of the adhesive under sustained loads, whereby the stiffening efficiency of the additional steel might be impaired. The upper limit on flexural modulus is to limit stress concentrations arising from strain incompatibilities at changes of section. The minimum value of bulk shear strength ensures that the adhesive is at least as strong as the concrete to which it is to be bonded with an appropriate factor of safety. It also assists in minimising the creep by keeping the working shear stress in the adhesive layer, typically 0.5–1.0 N/mm², at a relatively low pro-portion of its ultimate strength.

In the case of bond strength, steel-to-steel joints are recommended in Table 6.2 for determining fracture toughness and static and fatigue strengths. This is because steel-to-concrete joints are usually dependent on the concrete and as such provide no measure of the bonding efficiency with the external steel plate. The environmental conditions during operation and the required length of service for civil engineering structures need to be taken carefully into consideration when selecting an adhesive. A market evaluation sponsored by the Scottish Development Agency(19) suggests that the minimum required life for this strengthening technique applied to concrete bridges is 30 years. Clearly any accelerated laboratory tests selected, employing for example the wedge cleavage test, should demonstrate, as far as is practically possible that joints made with the chosen combination of surface pretreatment, primer and adhesive can survive the wide range of temperatures and condensing humidities likely at bridge sites whilst also subject to spray from de-icing salts, or, in the case of maritime structures, from the sea.

Table 6.2. Recommended mechanical properties of adhesive and adhesive joints for strengthening concrete structures (Ref. 18)

Property	Environmental condition for test	Recommended value	Test method
(1) Hardened adhesive			
Equilibrium water content	distilled water at 20 °C	max of 3% by weight	immerse 1 mm thick film of adhesive
Heat distortion temperature	within temp controlled cabinet	min of 40 °C	flexural test
Flexural modulus	20 °C	2000–10000 N/mm^2	flexural test
Shear strength	20 °C	min of 12 N/mm^2	shear box test
(2) Steel-to-steel joints			
Mode 1 fracture toughness	20 °C	min of 0.5 $MN/m^{-3/2}$	wedge cleavage specimen
Shear strength (static)	−25 °C to +45 °C	min of 8 N/mm^2	double overlap joint
Shear strength (fatigue)	20 °C	To survive 10^6 cycles of stress range between 0.4 and 4.0 N/mm^2.	double overlap joint

Surface preparation and curing. The concrete surface to be bonded must be sound, uncontaminated and free from chlorides. Before preparation any cracks wider than 0.3 mm and liable to leakage should be filled by injection of a suitable low viscosity resin. The existing surface must be roughened, using grit- or sandblasting, scabbling or a needle gun, to remove any weak material, surface laitance or contaminated concrete. Prior to applying the adhesive the prepared surface must be dry and free from dust. The surface of the steel to be bonded must also be free of contaminants including mill-scale, rust and most importantly, grease or oil. The degreasing and roughening procedure outlined in Table 6.3 is suitable(19),(20). A final solvent degrease after gritblasting is sometimes recommended but in the authors' experience this can redistribute contamination and do more harm than good (e.g. Fig. 3.9); certainly great care must be taken to ensure that there is an adequate time interval for solvent products to evaporate prior to adhesive spreading. Although the application of an epoxy based anti-corrosion primer to the prepared steel surface is often considered prudent to minimise any risk of subsequent corrosion, advice must be obtained to ensure that it is compatible with the adhesive system chosen or there may be a risk of interfacial bond breakdown. If there is likely to be any significant lapse of time, say greater than 24 hours, between the application of the primer and the adhesive, a primed steel surface should be given a final degrease before application of the adhesive itself.

Thorough mixing of the packaged and measured adhesive com-

Table 6.3. *Steel surface preparation procedure*

(1)	If necessary remove heavy layers of rust by hand or mechanical abrasion with emery cloth or by wire brushing to give rust grades A or B as defined in Swedish Standard SIS 055900 (20).
(2)	Degrease by brushing with a suitable solvent, e.g. acetone or 1.1.1 trichloroethylene, and allow to evaporate.
(3)	Grit blast to grade $2\frac{1}{2}$ of Swedish Standard SIS 055900 (20) to achieve a maximum peak-to-valley depth of at least 50 μm, using a hard angular clean metal grit which is free of any grease contamination.
(4)	Remove surface dust by brushing, vacuuming or blowing with a clean uncontaminated air supply.

Note: Step (4) should be followed as soon as possible by application of the adhesive.

ponents is carried out either by hand or with a slow speed mechanical mixer to avoid trapping air bubbles. The adhesive is then spread by hand in a thin layer to both steel and concrete surfaces, before the two parts are brought together and wedged or bolted into position. The final adhesive thickness should not be less than 1 mm although in practice the actual value will vary in excess of this depending on the flatness of the concrete surface to which the steel is bonded. During the curing period the ambient temperature needs to be maintained at a level of at least 10 °C for 24 hours with most formulations. Thus, it may be necessary to provide some form of external heating for winter repairs in the UK. Any temporary supports should be left in place until tests on control samples stored under the same conditions as the actual joint indicate that the required strength has been attained. In general there appears to be no need to stop normal traffic from using strengthened bridges while the bonding operations take place or during the curing period. Finally, the exposed steel surfaces should be painted with an anti-corrosion paint. This is ideally lapped over the exposed edge of the hardened adhesive and onto the concrete surface to minimise the possibility of moisture ingress along the steel/adhesive interface.

Adhesive quality control. Routine quality control tests on site are likely to be less comprehensive than the full scale materials testing programme which will be necessary for initial adhesive selection. In the former case check tests on pot-life and open time should be carried out on each batch of adhesive to ensure consistent results. Other routine tests suggested include four-point bending on samples of hardened adhesive (Fig. 6.14), joint shear strength using steel double lap specimens (Fig. 6.15) and pull-off adhesion on the concrete surface of the structure. It is also advisable to manufacture additional specimens for storage on site and testing at a later date

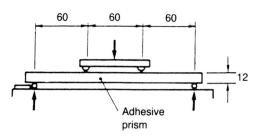

Fig. 6.14. 'All adhesive' flexural test.

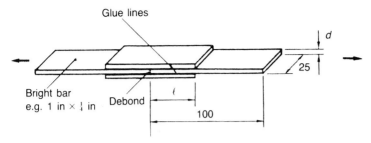

Fig. 6.15. Steel/steel double lap joint.

in order to assess long-term durability.

Tests to measure the bond which can be obtained with the concrete of the structure to be strengthened are best carried out on the structure itself. A possibility is to utilise a pull-off test as developed for the non-destructive testing of concrete(21). A circular steel probe is bonded to the concrete surface and specially designed portable apparatus is then used to pull off the probe, along with a bonded mass of concrete, by applying a direct tensile force. Any defects in bond would be revealed by the occurrence of failures at the adhesive–concrete interface.

6.3   Steel structures

Despite the extensive use of adhesives to bond structural metal in both the aircraft and vehicle industries for many years their use in civil engineering has been relatively rare. Early writers believed that resin adhesives could eventually replace welding in secondary connections in structural steelwork(22). For example, welding stiffeners to girder webs or diaphragms to prevent buckling failure often raises problems of distortion and fatigue. In the aerospace industry it is now established practice to achieve similar stiffening by bonding honeycomb structures with adhesives. The advent of toughened structural adhesives brought the opportunity for replacing welding in box beam metal structures such as used in the construction of motor vehicles(23).

Studies into the use of adhesives in the fabrication of metal bridge parapets have been called for because it is difficult to obtain adequate welded connections between lengths of horizontal rail. The rails are generally rectangular hollow sections and for a welded joint different electrodes have to be used for the top and bottom

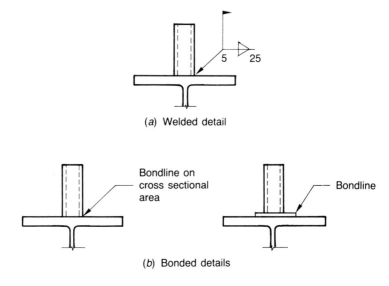

(a) Welded detail

(b) Bonded details

Fig. 6.16. Supports for concrete deck finishing machines (Ref. 24).

welds. Further, after completion the welds are ground flush to provide a smooth profile and a slight undercut can remove the weld completely. A simple bonded joint with either a splice or a sleeve could be designed to develop the full strength of the metal section.

In the United States some further possible applications were postulated in the form of bonded secondary steelwork connections on bridges(24). These included supports for concrete deck finishing machines (Fig. 6.16), sign support brackets (Fig. 6.17), lateral bracing connections (Fig. 6.18), drainage pipe supports (Fig. 6.19) and once again intermediate stiffeners to girder webs (Fig. 6.20). Experimental work had shown that, as a method for improving the fatigue life of bridges, adhesive bonds merited further investigation. However, it was carefully pointed out that a fundamental distinction needed to be made between short- and long-term projected uses. For example, the durability requirements for temporary construction attachments to girders are quite different from those for more permanent attachments such as vertical stiffeners.

For applications in new bridge construction a design life of 120 years is required(25). Although it may be possible to design out fatigue and creep by keeping working stresses low(26) the influence of the environment on adhesives over such a time span is very much

Steel structures

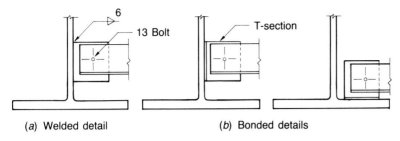

(a) Welded detail   (b) Bonded details

Fig. 6.17. Sign support brackets (Ref. 24).

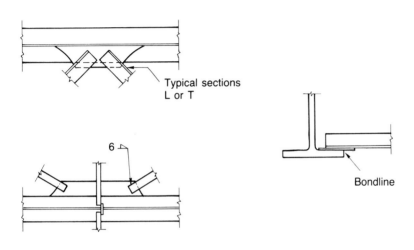

(a) Welded details   (b) Bonded detail

Fig. 6.18. Lateral bracing connections (Ref. 24).

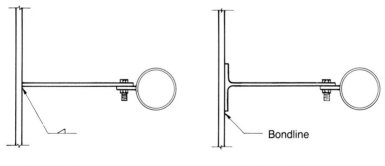

(a) Welded detail   (b) Bonded detail

Fig. 6.19. Drainage pipe supports (Ref. 24).

233

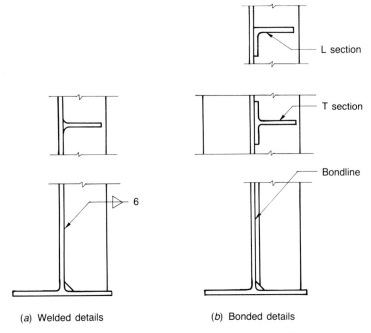

(a) Welded details   (b) Bonded details

Fig. 6.20. Girder web stiffeners (Ref. 24).

an unknown. It is for this reason that the application of adhesives in the repair and strengthening of steel structures is an area of more fruitful development. Design lives can be shorter, for example 30 years, in the knowledge that the bonded component may be replaced should deterioration occur. Indeed the Standing Committee on Structural Safety of the Institutions of Civil and Structural Engineers has consistently highlighted the lack of information on the long term performance of resins(27). They have warned against using resins in critical parts of a structure unless effective replacement can take place should structural deterioration occur.

One relatively recent application has been the bonding of existing Hobson's patent steel floor arches to replacement steel edge beams on the Branbridges bridge over the river Medway in Kent(28). An adhesive joint was selected in favour of welding or riveting because of restricted working room, speed and economy. A two part epoxy adhesive was injected into the space between a steel trimmer plate

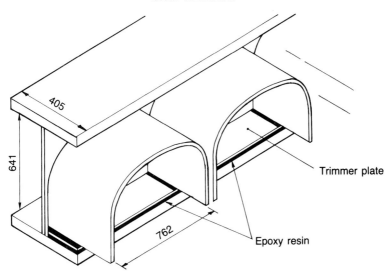

Fig. 6.21. Hobson's patent floor arches and trimmer plate bonded to new I beam (Ref. 28).

and the flange of the new edge beams (Fig. 6.21). A thixotropic filled epoxy was used to seal the gaps between the steel section prior to injection. The specification only allowed injection into the void between steel surfaces if their temperature was more than 3 °C above the dew point and the ambient temperature was in excess of 10 °C. To achieve this the contractor elected to sling a propane gas blower heater below the deck bridge.

The driving force for a change from traditional welding to adhesive bonding is thus the opportunity to avoid residual stresses. Possible applications that have been cited include large stiffened plate structures for both land and marine based industries(29). Such structures are designed to carry low to moderate forces mainly in shear and compression but may have to provide a substantial ultimate load carrying capacity as well as resistance to fatigue loading in service.

In Chapter 8 two potential applications for the bonding of structural steelwork will be considered in some detail. These are bonded web stiffeners and tension splices or cover plates to girder flanges.

6.4 Timber structures

There are two basic steps involved in the repair of a timber member. The first is to remove the reason for deterioration, for example by eradicating beetle attack or preventing the water penetration causing dry rot. The second is to select an appropriate repair technique and among the options available is the replacement of defective or rotted timber with reinforced synthetic mortar. In such cases the overall objective is to restore the original strength and load-bearing properties of a member. For upgrading or providing additional load-carrying capacity to existing sound timber, bolting, bonding, the insertion of new timber, steel plates, reinforcement bars or laminated beams are amongst the options available. The difference between timber and concrete repairs is that with timber renovation the synthetic resin is not the ultimate criterion in effecting a successful repair. Steel reinforcing bars or plates also play a major role in the total structural system(30).

Cracks and shakes in the timber may be initially consolidated by the injection of synthetic resins. This serves to re-establish the member's ability to support its own weight and provides a suitable base for bonding subsequently applied repair mortars. Epoxy, polyester or acrylic resins can be suitably formulated to provide the fluidity necessary for maximum penetration and impregnation of the wood fibres. Introduction of a solvent just prior to use enables the resin to be soaked up by the timber before chemical hardening occurs. Any decayed timber is then replaced with a resin mortar, the strength and properties of which may be varied to allow subsequent working, for example nailing or sanding. Structural stability is restored by introducing steel reinforcement elements into the damaged area, for example as Fig. 6.22 shows for the rejoining of a rotted mid-section of a beam.

In old buildings timber often fails where it enters a masonry wall or in roofs at the junction of rafter and tie beam where the truss is supported on the wall plate. In both cases leaking water over a long period has resulted in wet rot. The repair process consists of replacing the decayed timber with reinforced epoxy mortar. The reinforcement may consist of steel plate or bars which are drilled or set deep into the sound portion of the timber to ensure a good bond and load transfer. Fig. 6.23 shows bonded dowel bars used to repair the end section of a column or beam while Fig. 6.24 illustrates a typical repair to a roof truss.

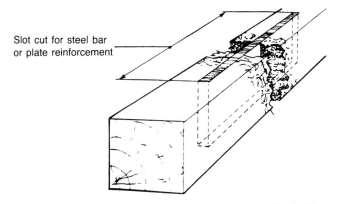

Slot cut for steel bar
or plate reinforcement

Fig. 6.22. Rejoining mid-section of timber beam (Ref. 30).

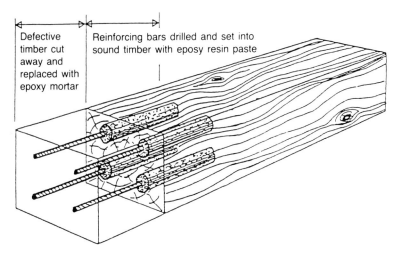

Defective
timber cut
away and
replaced with
epoxy mortar

Reinforcing bars drilled and set into
sound timber with eposy resin paste

Fig. 6.23. End repair to timber section (courtesy Resin Bonded Repairs Ltd).

Deteriorated or worn timber floors can be upgraded by bonding new timber to old with epoxy resin. The procedure involves removing all deteriorated timber back to a sound surface prior to the application of a penetrating epoxy sealer. The new timber is then bonded to the old and composite action ensured by drilling and inserting dowel pins set in resin as shown in Fig. 6.25.

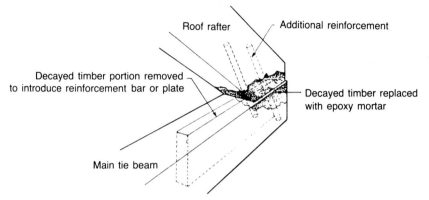

Fig. 6.24. Repair of decayed roof truss (Ref. 30).

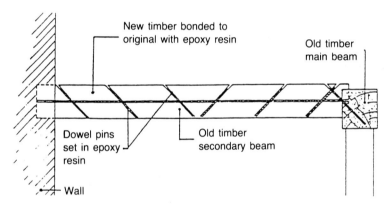

Fig. 6.25. Upgrading a timber floor (courtesy Resin Bonded Repairs Ltd).

6.5 Masonry structures

Apart from the use of resin adhesives to effect repairs to failed brick slips, perhaps the three most common applications in the repair and/or strengthening of masonry structures involve injection, the fixing of dowels and the use of bedding mortars. Resin injection is often used to stabilise deteriorating brickwork in listed structures. It involves drilling through the exposed mortar joints, setting in injection tubes and pumping in low viscosity resin (polyester or

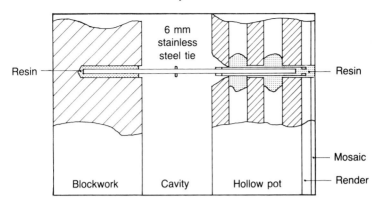

Fig. 6.26. Cavity wall tie (Ref. 31).

epoxy) to consolidate loose rubble infill and to seal all cracks and voids in the brickwork itself.

There are a number of proprietary systems for resin bonded anchors which may be used to install additional ties in cavity walls. A typical fixing detail is shown in Fig. 6.26 which is based on a 6 mm stainless steel tie(31). This is bonded into pre-drilled holes in the hollow or solid masonry using resins formulated to suit the type of base material. For voided materials the resins may be injected via a mesh sieve or screen which enables a keying action to be obtained. Fig. 6.27 illustrates the steps to be followed in the fixing procedure. Similar details may be used for fixing steel anchors into unreinforced brick walls for improved strength and stability. Two-component, unsaturated polyester resins contained in separate cartridges which are subsequently mixed in a combined dispensing and mixing nozzle are commonly encountered. Further details of systems available may be found in Chapter 7 dealing with applications in new construction.

Applications are not necessarily confined to buildings. Nineteenth century masonry arch bridges have suffered damage and deterioration in recent years as a result of having to carry modern traffic loads. Defects in the form of longitudinal arch cracking and spalling of voussoir stones have been reported(32). Fig. 6.28 illustrates several uses of adhesives in a major refurbishment contract. 40 mm diameter stainless steel tie bars of length 1.5 m and bonded in resin were used to tie the edges of the arch barrels back to the centre.

1. Drill a 7/16″ pilot hole and enlarge the hole with a 1″ bit.

4. Insert the filled screen tube into the hole.

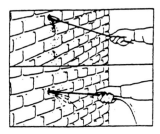

2. Clean the hole with a nylon brush and compressed air.

5. Push the threaded rod into the filled screen tube.

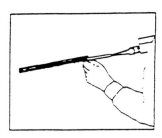

3. Beginning at the bottom, fill the screen tube with adhesive.

Fig. 6.27. Fixing of resin bonded anchors to masonry (courtesy of Hilti Ltd).

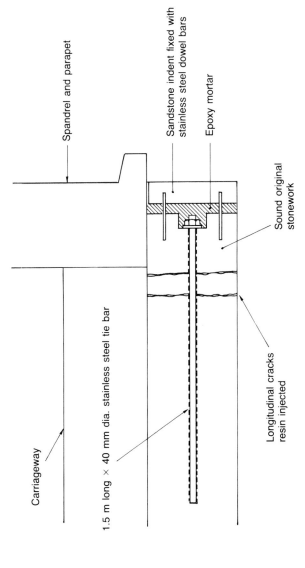

Spandrel and parapet

Sandstone indent fixed with stainless steel dowel bars

Epoxy mortar

Sound original stonework

Carriageway

1.5 m long × 40 mm dia. stainless steel tie bar

Longitudinal cracks resin injected

Fig. 6.28. Refurbishment of a masonry arch bridge (Ref. 32).

Longitudinal cracks were first filled by injection of a variety of materials depending on the crack width. Large cracks in excess of 15 mm width were injected using cement grout containing a non-shrink additive. A filled epoxy resin was used for cracks ranging from 1 to 15 mm wide whilst a low viscosity epoxy was used for the finer cracks. Replacement surface stonework was bedded on epoxy mortar and tied back into sound original stonework with 16 mm stainless steel dowel bars anchored using a very fast setting resin.

CHAPTER SEVEN

Applications in new construction

7.1 Bearings and expansion joints for bridges

Perhaps the most well known application of resins in civil engineering is in the form of resin mortar for either bridge bearings or expansion joint nosings. In bearings the mortar is used as a bedding compound on which to seat rubber or steel bearing pads. These pads serve to transfer loads from the superstructure to piers or abutments and the stresses they resist are largely compressive in nature.

Expansion joints in concrete bridges were traditionally formed by the use of steel edging angles anchored or bolted into the concrete deck either side of the expansion gap. Such joints have a limited life due to disintegration of the supporting concrete under the action of traffic impact and they are very difficult to repair or replace. As a result resin mortar nosings were introduced on UK highway bridges in the mid 1960s but they too were not entirely free from trouble. The early nosings were based on relatively slow curing epoxy systems and a typical installation procedure is illustrated in Fig. 7.1(1). In due course other epoxy formulations began to appear and faster curing systems were introduced in an attempt to avoid the need for heating and testing in cold weather.

In the late 1960s transverse cracks began to appear in some nosings and in the following years hollowness and cracking became a major nuisance. Investigations revealed that the majority of faults were due to use of the faster curing systems, resulting in warping and curling as shown in Fig. 7.2. This occurs because of differential contraction following the exothermic reaction, the bottom of the nosing being restrained by adhesion to the concrete deck. On cooling the nosing curls and lifts at the transverse joints or shrinkage cracks, and may be accompanied by cracks extending down into the concrete deck.

In more recent years such difficulties have largely been overcome by careful formulation of the resin. This includes the use of resins with a slower cure rate, the use of larger size aggregate to allow

243

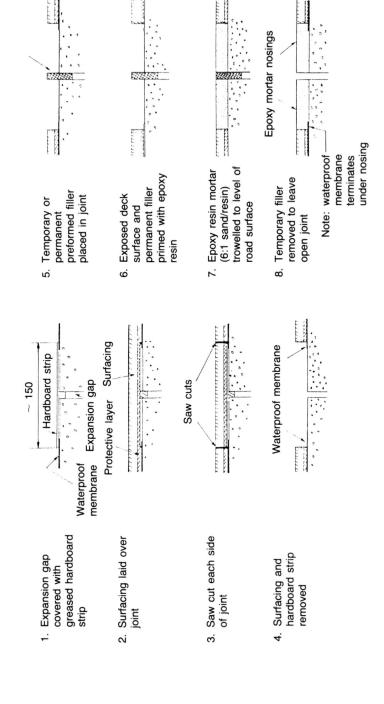

1. Expansion gap covered with greased hardboard strip

~ 150

Hardboard strip

Waterproof membrane

Expansion gap

Protective layer   Surfacing

2. Surfacing laid over joint

Saw cuts

3. Saw cut each side of joint

Waterproof membrane

4. Surfacing and hardboard strip removed

5. Temporary or permanent preformed filler placed in joint

6. Exposed deck surface and permanent filler primed with epoxy resin

7. Epoxy resin mortar (6:1 sand/resin) trowelled to level of road surface

8. Temporary filler removed to leave open joint

Epoxy mortar nosings

Note: waterproof membrane terminates under nosing

Fig. 7.1. Installation of epoxy nosing (Ref. 1).

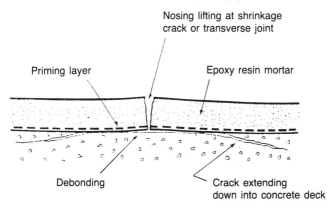

Fig. 7.2. Curling of expansion joint nosing (Ref. 1).

heat to dissipate or by flexibilising the epoxy to permit some stress relief during cure. Polyester resins have also been considered and indeed used on one occasion. They have the attraction of their cure being less affected by low temperatures but the disadvantage of higher shrinkage and poor adhesion to damp surfaces. Polyurethane resin mortars have also been used successfully for a limited number of applications. They are cheaper than epoxies and can cure at lower temperatures but have a tendency to foam in the presence of water.

Despite the development of resin mortar technology in recent years such that many of the causes of the earlier failures have now been eliminated, the confidence of the industry has been severely dented and resin nosings are now a rarity in UK bridge construction.

7.2   Skid resistant surfacings

Some of the earliest major applications of adhesives in civil engineering involved the use of resins for abrasion resistant and non-slip surfaces to heavy duty floors and roads. This was achieved by the use of synthetic anti-skid grits, such as calcined bauxite, set in a resin base. Both epoxy and polyester resins have been used, applied either by trowel or in slurry form by squeegee. Trowelled systems are usually heavily filled mortars with an aggregate:resin ratio of the order of 6 : 1 and as such usually require a priming coat on both steel and concrete surfaces. In contrast resin-rich slurry systems are self-wetting on steel surfaces but may require a primer when applied to concrete. Like mortars they can be gritted after application but the grit must only partially sink into the slurry to give an appropriate degree of embedment for both adhesion and skid resistance.

The initial interest in improving skid resistance stems from the fact that a high proportion of road accidents involving casualties occur within a short distance of road junctions and pedestrian crossings(2). Many of the road surfaces at such locations have poor skid resistance and this led the Transport and Road Research Laboratory to conduct trials into methods of resurfacing which did not eventually polish under traffic(3). Small-scale trials were initiated in London in 1966 using epoxy based adhesives and these were so promising that the then Greater London Council gave the go ahead for full-scale trials at seven major road junctions the following year. The success of these trials has since led to the development of proprietary systems for treating asphaltic road surfaces which are accompanied by automated metering, mixing and application systems capable of applying the epoxy at a rate of up to 80 m² per minute(4).

Resin-based surfacings have also played a part in temporary structures and some interesting trials were carried out in 1970 on a flyover in East London(1). The structure had a precast concrete deck and a number of steel removable panels in the approach ramp carriageways. About twenty different surfacing systems were applied to the panels, including polymer cement, polyurethane, polyester and epoxy resins. Ten years later three of the epoxy systems and one polymer cement material were still found to be in good condition. However, the repair of failed panels in the intervening period created problems due to weather conditions and short overnight closure periods. This precluded epoxy systems and a polyester resin mortar was used. In service the polyester was found to be less wear resistant than the more successful epoxy compounds and after 12 months trafficking it would lose grit and polish.

On steel surfaces solvent degreasing and gritblasting is recommended as the surface preparation method, although where this has to be done in the factory prior to delivery, application of an anti-corrosion priming layer is often specified. A primer which is compatible with the surfacing formulation must be used but, even when applied correctly, there is usually some reduction in bond strength. If a primer is used it is recommended that all solvents be allowed to evaporate prior to application of the resin itself. On concrete surfaces, where all laitance must be removed and the surface be sound, dry and free from oil, grease and dust, a wetting coat of unfilled binder may be necessary immediately prior to final surfacing. Such principles are no different from those described in earlier chapters for efficient bonding to steel and concrete substrates in other applications.

Nowadays resin bonded anti-slip surfaces are a common feature of many pedestrian walkways including footbridges. Given appropriate substrate preparation and careful selection of the adhesive they can prove to be a lightweight, durable and effective surfacing method.

7.3   Resin bonded fixings to concrete, masonry and rocks

There are two basic methods by which steel bolts and bars may be anchored into concrete, masonry or brickwork utilising epoxy or polyester resin grouts(5). The first involves inserting a deformed bar or bolt into a preformed or drilled hole into which the resin grout has been poured (Fig. 7.3).

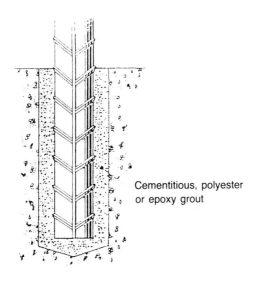

Cementitious, polyester or epoxy grout

Fig. 7.3. Fixing using pourable resins (Ref. 5).

247

Table 7.1. *Typical resin quantities for fixings using pourable resins (ml per 100 mm of bonded length)*

Hole diameter (mm)	Bolt Diameter (mm)				
	12	16	20	25	32
20	25				
25	50	40	25		
32	80	70	60	40	
38		100	100	75	45
45			150	130	100
51				180	150
57				235	200
64				300	270
76					425

Maximum anchorage strengths and efficiency are achieved in holes drilled with rotary percussive air drills using an air or water flush to remove dust. Diamond drilled holes are not recommended unless the sides are roughened or are under-reamed using special tools. After the holes have been drilled the correct volume of resin should be mixed, according to the manufacturer's instructions, and poured in. Table 7.1 typifies the information provided by manufacturers of proprietary systems to assist in estimating the quantity of resin to be used. The bar or bolt is then lowered into the hole and pressed gently to the bottom, displacing the resin to grout the annulus around the bolt. A slight agitation of the bolt may assist the grout to settle evenly around the anchorage, especially in cold weather. When the bolt is centred it should be left undisturbed until the resin is set.

The failure mode of grouted bars may be by conical rupture of the base material or slip at the resin/substrate interface. Edge distances and centres must be sufficient to generate the full pull-out strength which is related to the bolt type and diameter, the embedment length and the strength of the substrate. The design procedure is to select the bolt or bar diameter according to the load to be carried and the characteristic strength of the deformed portion. The embedment length, L, is then calculated using relationships of the form $L = (50 + 25P)$ mm where P is the load to be carried in tonnes multiplied by an appropriate factor of safety(6). A factor of 2 is normally applied. Potential performance may be affected by:

(1)   inadequate cleaning of holes after drilling
(2)   low or high temperatures
(3)   fire
(4)   creep under sustained load
(5)   cyclic or vibrating loads.

The second type of system is marketed as encapsulated packs of polyester or epoxy resin within which there is a thin skin to keep the two components separated. The capsules are placed in the hole and are fractured when the bolt or bar is inserted and turned (Fig. 7.4). In all such systems it is essential that the fixing is mechanically rotated in the hole to ensure correct mixing of the resin components. The requirements for drilling and cleaning the hole are similar to the poured grout systems but the hole size should be strictly in accordance with the manufacturer's instructions. This is to ensure that the pre-measured volume of resin in the pack completely fills the annulus between the bolt and the hole.

The failure modes and the performance of encapsulated resin fixings are similar to those described for the grouted systems. Manufacturers frequently provide graphs of the form shown in Fig. 7.5 for determining the required anchor length. In using such graphs it is important to ensure that an adequate bolt type and size is first selected, that the graph is applicable to the substrate material and that the load is multiplied by an appropriate factor of safety prior to use.

The concept of encapsulated packets of resin has proved most popular for polyester resin grouts. The degree of intimate mixing required is not so critical as with epoxies, curing rates are faster and installation can take place at lower temperatures. The range of applications is enormous – the fixing of reinforcement starter bars, foundation bolts, machinery and base plates, barriers and safety fences, railway and crane rails, etc.

Resin bonded systems employing the larger bar diameters and lengths of several metres have also proved popular for rock bolting and ground anchor systems for example as shown in Fig. 7.6. Three basic systems are available:

(1)   The inner end of the bolt is anchored with a resin capsule and the bolt shank subsequently tensioned against a bearing plate on the rock surface. A compressive force is thus applied to counteract forces acting to expand the rock mass. The tensioned bolt may be fitted with load indicating devices to enable any

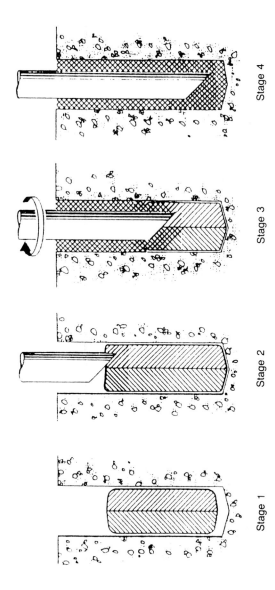

Stage 1 Stage 2 Stage 3 Stage 4

Fig. 7.4. Fixings using encapsulated resins (Ref. 5).

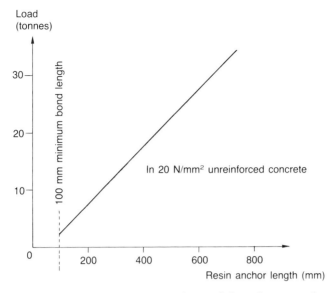

Fig. 7.5. Load *v* anchor length for fixings in unreinforced concrete (courtesy of CBP Ltd).

Fig. 7.6. Rock bolting in highway cutting.

ground movement to be monitored. Fig. 7.7 shows typical end bonded lengths against maximum load for several rock types.

(2) A bolt which may be a smooth or deformed bar is bonded over its complete length. The bolt is untensioned but since the hole is completely filled with cured resin it can offer good resistance to shear stresses. Spalling of the surface rock, which can render tensioned bolts ineffective, does not impair the performance of a fully bonded dowel. Further, the resin can act to provide corrosion resistance to the bolt as well as providing the necessary bond.

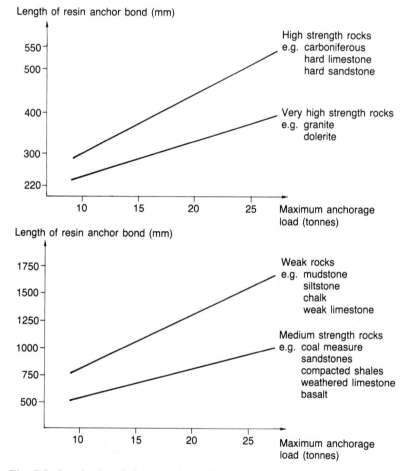

Fig. 7.7. Load v bonded length for end bonded rock anchors (courtesy of Celtite (Selfix) Ltd).

(3) The third system combines the merits of both the tensioned bolt and the fully bonded dowel. A fast setting resin is used for the end anchorage, thus enabling tensioning of the bolt before the slower setting resin, used over the remainder of the length, has gelled. The method combines permanent reinforcement of the ground with positive compression to minimise movement at potential slip planes within the rock strata.

7.4   Wire and strand anchors

Steel wire and strand formed into cables is commonly used in the prestressing of concrete and in suspended or cable stayed structures such as bridges. In both cases it is sometimes necessary to splay the constituent wires or strands at the end anchorages in order to spread the load being transferred to the support material.

In cable-supported bridges this is achieved by splaying the strand ends into a conical socket which is then filled with molten metal. When the cable is tensioned the conically formed deflector is pulled into the socket and load transferred in compression by wedge action. In some cases (1) the molten metal has been replaced by a low viscosity two-part resin, usually a filled polyester for its speed of cure. This obviates the need for heating and enables the attachment to be formed on site. Although it has been argued that the frictional force between the strands and the anchorage is sufficient to retain the cable end in the socket, some interfacial bond is necessary otherwise the strands would pull out before the wedging action could develop. Similar techniques are now being considered for use in sleeve couplers designed to connect steel bars for the reinforcement of concrete (see Chapter 8).

Pourable epoxy resins have also been used to bed deflector plates and anchorage plates in certain special applications of prestressed concrete(7). One such example was a farm access bridge over the northern end of the M1 motorway. Following the discovery of cracking in some parts of the concrete box which formed the arch structure a strengthening method was sought which would not detract from the appearance of the bridge. A system of prestressing within the box was evolved but this required anchorages capable of providing adequate strength to spread the load within the corners of the bottom of the box. Hence the use of deflector and anchor plates bedded in epoxy resin.

7.5 Composite steel–concrete construction

A potential structural application of epoxy resin adhesives in bridge construction is to use them to form the necessary shear connection between steel girders and the concrete deck slab in composite bridges, in place of conventional welded mechanical fasteners. There are three ways in which bonded shear connections between steel and concrete might be achieved.

Post-bonding. Adhesives may be used to bond precast concrete slab units directly to the steel surface. Such a technique has been successfully employed in Germany(8). On one bridge epoxy resins were used in a variety of forms and locations on the superstructure (see Fig. 7.8). This included rubber bearings bedded in epoxy mortar and an epoxy adhesive coating between the main steel girders and the bearings. Of most interest structurally, however, is the use of epoxy resin adhesive mortar between the precast concrete deck panels and the main girders. The intention was to create a monolithic structure having the same degree of strength and composite action as might be expected with conventional jointing methods. Four

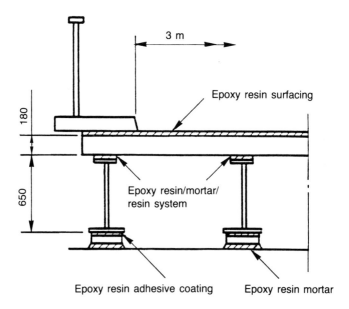

Fig. 7.8. German bonded bridge (Ref. 8).

years after construction the bridge was subjected to full scale load tests and the resulting deflections compared with those based on full and zero composite action. In both the load cases considered, actual deflections were never more than those calculated assuming full effective bonding and were four to five times less than in the unbonded case. It is also worth mentioning that such results were obtained after a period in which the river being bridged rose to the level of the superstructure on three occasions and, in one case, completely submerged the entire structure. It is reported that subsequent visual monitoring has revealed the joint edges to be free of cracks or other damage. The final use of adhesives in this structure was in the form of an epoxy adhesive surfacing on the precast concrete deck panels which was also employed to seat the precast concrete kerb units.

Grip layer. A layer of coarse aggregate may be bonded to the steel to form a rough layer onto which fresh concrete is subsequently poured. This technique has been the subject of some research involving slab panels, tensile reinforcement being provided in the form of a thin soffit steel plate. However, premature adhesive failures have been reported and an improved bonding technique is considered to be necessary(9).

Pre-bonding. Fresh concrete is poured directly onto a layer of uncured adhesive spread over the prepared steel surface. This method, as used in the reinforcement of slab units by externally bonded plates, has been the subject of extensive research by the Wolfson Bridge Research Unit(10). 'Open sandwich' slabs, based on either a flat soffit plate or a plate curved slightly upwards to form a shallow arch have been recommended. The technique is described in some detail in Chapter 8.

7.6   Segmental concrete construction

Applications of resins as true adhesives in new construction have been relatively scarce but increasing use is being made of epoxies in the joints between units in segmental precast, prestressed bridge construction (see Fig. 7.9). Using this technique long span bridges can be constructed by stressing precast concrete segments together to form a monolithic structure. The advantages of segmental

Precast concrete
box sections

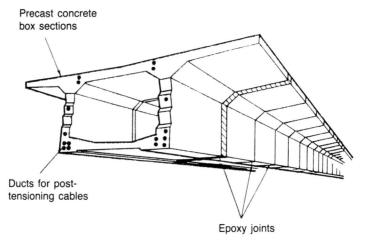

Ducts for post-
tensioning cables

Epoxy joints

Fig. 7.9. Segmental bridge construction.

construction are that it speeds erection, can accommodate changes in horizontal and vertical curvature, and the relatively small segments are relatively easy to handle and transport. Cost savings of the order of 20% have been claimed over alternative methods of construction(1).

Traditionally the joint zones had been formed of widths in excess of 250 mm, which were filled with cast *in situ* concrete, or as relatively narrow joints, 20–40 mm thick, filled with dry packed sand/cement mortar. Concrete joints need time to cure and as a result it was sometimes necessary to temporarily support several segments before stressing could take place. In 1963, on a bridge over the Seine at Choisy-le-Roi outside Paris the precast segments were counter cast for the first time. This involves casting each segment against the previous one whose mating surface has been precoated with a release agent. The objective is to produce a very thin accurate joint of width 1–2 mm which can subsequently be filled with an epoxy resin. Although the resins are much more expensive than concrete the use of the thinner joint offsets this cost. Further, the waiting time before the segments can be stressed together is considerably reduced and hence leads to a faster speed of erection. The adhesive also assists in erection by acting as a lubricant during final alignment. More recent examples of use of the technique in the UK include the Trent Bridge near Scunthorpe on the M 180 motorway(11) and the East Moors Viaduct, Cardiff(12).

The resin in the joint is not used in a truly structural sense since it is only designed as a gap filler to transmit compressive stresses. However, it does serve to more evenly distribute these stresses. The adhesives have not normally been used to resist the vertical shear which develops between adjacent units since keys are formed for this purpose, but there is no reason why they should not be designed to do so. The reluctance stems from the 120-year design life requirement for highway bridges in the UK and the lack of sufficient data on the long-term durability of adhesives. It is ironic, therefore, that one of the major claimed advantages of the technique is the relatively impermeable joint which is obtained compared with conventional cement mortars. The collapse of the Ynysgwas bridge in South Wales in 1985(13) bears testimony to the lack of resistance to moisture penetration of some wide sand/cement mortar joints.

A draft standard for acceptance tests and verification of epoxy bonding agents for segmental construction was published in 1978(14). The need arose because of the perceived difficulties of using epoxies for segmental work under conditions of varying humidity and temperature. The functions of the bonding agent are described as:

(1)   To join the surfaces of the precast concrete segments in such a way that compression and shear, and in some cases also tensile stresses, are transmitted between the segments.
(2)   To develop such strength, at a rate sufficient to allow continuous erection of the segments.
(3)   To lubricate the surfaces of the joints to facilitate proper positioning of the segments during erection.
(4)   To provide a moisture seal across the joints to protect the prestressing tendons against corrosion and to prevent leakage at joints during tendon grouting.

To fulfil these requirements the proposal suggests separate programmes of testing for (a) choosing an appropriate bonding agent and (b) site testing. The properties required and the proposed test methods are summarised in Tables 7.2 and 7.3.

Epoxy adhesives were used in 1978 during the construction of the U.S.'s first segmentally constructed cable-stayed bridge across the Columbia River(15). Here the deck is made monolithic by a combination of post-tensioning and compressive forces from the horizontal components of the stay cable forces; the epoxy mortar is called upon to provide both shear and tensile resistance across the joint.

257

Table 7.2. *Programme of testing when choosing a bonding agent for segmental construction* (Ref. 14)

Property	Test Method
Pot-life	temperature change of insulated cylinders
Open time	series of tensile bend tests with time
Thixotropy	Daniels gauge or sag flow board
Angle of internal friction	squeezability between flat plates
Bond to concrete	tensile bend test on bonded prisms
Curing rate	compressive strength with time
Shrinkage	length change of prisms
Creep	deferred compression and shear moduli
Water absorption	weight change of rods
Heat resistance	DIN 53457 (Martens) on rods
Colour	similarity to concrete
Compressive strength	unspecified
Modulus in compression	prism under uniaxial compression
Tensile bending strength	4-point bend test on bonded prisms
Shear strength	slant prisms or cylinder
Shear modulus	torque of cylinders

Table 7.3. *Site testing of the bonding agent for segmental construction* (Ref. 14)

Property	Test Method
Seviceability	appropriate storage and checks for crystallisation of resin.
Pot life	change of temperature of the mix with time using a thermocouple.
Open time	manufacture of lap joints at regular time intervals using asbestos board and hand testing
Colour	similarity with concrete.
Rate of curing	if possible use compressive or tensile bend strength as in Table 7.2. Alternatives include bonding of concrete test specimens to a segment in place or bonding of small steel cylinders to concrete test cubes.

Three adhesive formulations were chosen to cover the anticipated temperature range at the time of application, i.e. 4 °C–38 °C. Control of correct mixing and materials was achieved by measuring hardness of small specimens after curing for 20 minutes in a small oven. Subsequently slant shear and compressive strength tests were performed.

Construction of the M 180 bridge near Scunthorpe (Fig. 7.10) is particularly interesting since an aliphatic amine cured epoxy was used during severe UK winter conditions. A technical account of the site-testing carried out describes the procedures used to examine the curing characteristics of the adhesive in cold weather(16). Cubes bonded together to form a beam and subsequently tested for flexural strength, deflection and creep revealed little difference from monolithic concrete provided the concrete surface to be joined had been properly prepared. A minimum compressive stress of 0.3 N/mm^2 is also recommended during the curing period. Single lap joints were used to monitor the effect of curing temperatures on subsequent strength under both site and controlled laboratory conditions. The results shown in Fig. 7.11 for site joints relate to the maximum temperature of the range experienced. They suggest that for the particular adhesive, the lowest reliable limit for full curing must be

Fig. 7.10. M180 bridge.

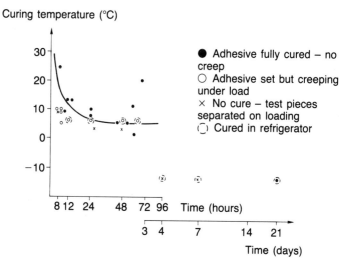

Fig. 7.11. Effect of low temperature on cure rate of an epoxy (Ref. 16).

at least 3 days at temperatures in excess of 5 °C. Below this temperature site heating must be employed.

Creep of the epoxy resin system under sustained load has always been of some concern. However, with a thin glue-line and at the relatively low stress levels encountered in segmental construction, compressive creep should not be a problem with most epoxies(2).

7.7 Epoxy coated rebars

The use of reinforced concrete as a construction material has expanded enormously since the 1960s, not least for road and bridge structures and for marine applications. In both situations there is the possibility of ingress of chloride ions, in the first instance from the use of de-icing salts and in the second from the effects of sea spray. The presence of chlorides in slightly moist concrete creates an electrolyte in which small galvanic currents can flow. Because of variations in conditions along the length of a reinforcement bar, electrical potential differences occur and currents flow from anode to cathode. Material is dissolved at the anodes resulting in either general or local corrosion of the reinforcement bars. The former manifests itself as red rust which is expansive so that cracking and

260

then spalling of the cover concrete takes place (see Fig. 6.1). Localised corrosion in the form of black rust occurs at discrete sites and causes a much greater loss in reinforcement cross-sectional area. However, the chemical conversion is less expansive so that there is little disruption to the concrete and as a consequence it is more difficult to detect.

The consequences of the liberal use of de-icing salts on unprotected concrete bridge decks in North America has been quite dramatic. General corrosion has occurred on many decks within the first 5–10 years necessitating expensive remedial repairs. Research in the United States in the early 1970s(17) concluded that organic coatings, particularly epoxies, could be used to protect steel reinforcement bars in the concrete of bridge decks and buildings from rapid corrosion. During 1973 the first highway bridge to use epoxy coated reinforcement was constructed in Pennsylvania and from 1978 electrostatic epoxy-powder coated reinforcement (EECR) became a standard construction material in the USA and Canada(18).

In the UK various techniques have been considered to make the reinforcement more resistant to corrosion. These include galvanised reinforcement and the use of stainless steel. Epoxy coated bars complying with the relevant ASTM standard(19) were first brought into the UK in the early 1980s. However, the wide variation in coating thickness revealed by microscopy was regarded as unacceptable for Europe(20). To meet the European requirements a process capable of producing a uniform thickness of coating, in the range 150–250 microns, has been developed. The product is known as fusion bonded epoxy coated rebar (FBECR).

The process involves three basic stages:

(1) Surface preparation. The surface of the rebar is blasted to a surface cleanliness at least as good as Swedish Standard ASa 2.5. A surface texture of around 70 μm depth is aimed for. This is typical of steel surface preparation requirements for other structural bonding applications (see Chapter 6).

(2) Heating. The rebars are then heated using an induction heater. This does not contaminate the already clean bar surface and allows a constant coil voltage combined with a constant bar speed to produce very uniform surface temperatures.

(3) Coating. Traditionally coating has been done by electrostatic spraying rather than by dipping in a fluid bed of epoxy. The simple spraying of charged particles is limited in terms of the

uniformity of coating which can be achieved on a deformed cross-section rebar. A technique known as tribostatic charging, whereby the particles of powder coating are charged by friction, offers advantages when coating rebar.

Europe's first rebar coating facility was opened in Cardiff in 1987. In comparison with North America where use in road bridge decks predominates, applications in the UK have been concentrated towards marine and water retaining structures. There are still some reservations about more widespread use based on the results of research by the Transport and Road Research Laboratory(21). These include reductions in bond performance for cold twisted bars, the effect of fatigue on the integrity of the coating, especially at deformations in the bar, and concern as to how defective areas can be repaired. The ability of epoxy coated bars to retain their integrity over long periods of time in alkaline environments is also questioned. Research at the Building Research Establishment(22) has shown that FBECR provides a significant reduction in the rate of deterioration of reinforced concrete containing chlorides. However, the use of these coatings does not provide total protection since corrosion may be initiated at breaks in the film. Nevertheless epoxy coating has been preferred to other methods for providing added protection to concrete bridges at the design stage(23).

7.8   Glued laminated timber

The adhesives to be used in making joints between timber members are specified in BS 5268: Part 2 (24). Permissible adhesive types based on the exposure conditions of the joint are reproduced here as Table 7.4. 'Weather and Boil Proof' (WBP) products such as resorcinol and phenol-formaldehyde are stipulated in damp conditions whereas melamine and urea-formaldehyde adhesives are permitted in permanently dry situations. Natural adhesives such as casein are now only allowed in the least hostile exposure category.

One particular structural application of adhesives is in the manufacture of glued laminated timber or 'glulam' members. This is normally manufactured by glueing together at least four planks of timber with their grain essentially parallel. The laminations are usually machined from 38 or 50 mm thick timber although thinner sections may be necessary in the production of curved members. End

Glued laminated timber

Fig. 7.12. Glulam hardwood footbridge.

jointing to obtain full length laminations is usually obtained by using finger joints.

Adhesives used in the construction of glulam are required to have a strength sufficient to provide a joint at least as strong as that of the timber in shear parallel to the grain. This implies a value normally not greater than 1 N/mm^2 for most softwoods. Careful selection of adhesive type in accordance with the guidelines of Table 7.4 is necessary for longer term applications, particularly in damp conditions or outdoors. Fig. 7.12 shows the use of laminated hardwood for the main girders of a footbridge.

Glue spreading and clamping must be accurately and rapidly carried out as adhesives have limited pot-life and assembly times(26). To achieve a controlled spread of glue, laminates are coated using double roller glue spreaders, and jigs are used to assemble the glued laminations. They are then clamped under controlled pressure to achieve a thin glue-line, the clamping force being maintained by hydraulic, pneumatic or mechanical means. The clamped laminations are held at a steady temperature until the adhesive is fully cured, the curing time depending upon the type of adhesive and the temperature applied. After completion of the curing the member is conditioned for a period at room temperature before it is put into service.

Table 7.4. *Permissible adhesive types (Ref. 24)*

Exposure	Category	Typical exposure conditions	Adhesive type		BS reference
Exterior					
	High hazard	Full exposure to the weather, e.g. marine structures and exterior structures where the glue-line is exposed to the elements. Glued structures other than glued laminated members are not recommended for use under this exposure condition.	Resorcinol-formaldyehyde (RF) Phenol-formaldehyde (PF) Phenol/resorcinol formaldehyde (PF/RF)	} Type WBP	BS 1204: Part 1 (25)
	Low hazard	Protected from sun and rain, roofs of open sheds and porches. Temporary structures such as concrete formwork.	Resorcinol-formaldehyde (RF) Phenol-formaldehyde (PF) Phenol/resorcinol formaldehyde (PF/RF) Melamine/urea formaldehyde (MF/UF)* Urea-formaldehyde (UF)* Other modified Urea-formaldehyde (UF)*	} Type WBP Type BR Type MR Type BR	BS 1204: Part 1 (25)

Interior

	Location/Conditions	Adhesive	Type	
High hazard	Building with warm and damp conditions where a moisture content of 17% is exceeded and where the glue-line temperature can exceed 50 °C, e.g. laundries and unventilated roof spaces. Chemically polluted atmospheres, e.g. chemical works, dye works and swimming pools. External single-leaf walls with protective cladding.	Resorcinol-formaldyehyde (RF) Phenol-formaldehyde (PF) Phenol/resorcinol formaldehyde (PF/RF)	} Type WBP	BS 1204: Part 1 (25)
Low hazard	Heated and ventilated buildings where the mosture content of the wood will npt exceed 17% and where the temperature of the glue-line will remain below 50 °C, e.g. interior of houses halls, churches and other buildings.	Resorcinol-formaldehyde (RF) Phenol-formaldehyde (PF) Phenol/resorcinol formaldehyde (PF/RF) Melamine/urea formaldehyde (MF/UF)* Other modified Urea-formaldehyde (UF)* Urea-formaldehyde (UF)*	} Type WPB Type BR Type BR Type MR	BS 1204: Part 1 (25)
	Inner leaf of cavity walls	Casein		previously specified in BS1444 which is now withdrawn

*The designer should ensure that a particular formulation is suitable for the service conditions and for the intended life of the structure.

Figure 7.13 Curved glulam portal frames.

A major advantage of glulam arises from the use of graded laminates in the reconstitution process. This causes a reduction in the natural variability of the timber, thereby permitting higher design stresses than in an equivalent solid member. Further, the flexible members can be held in a bent form during the curing process allowing graceful curved members as shown in Fig. 7.13 to be produced. These technical advantages combined with high fire resistance and attractive appearance suggest an increased application in the future, particularly where clients are prepared to pay a little extra for improved aesthetics. Costs are likely to reduce now that some manufacturers maintain stocks of standard straight beams which are available for purchase 'off the shelf'.

CHAPTER EIGHT
Potential future developments

8.1 Introduction

It has already been observed that structural adhesive bonding, either alone or in combination with other methods of fastening, represents a key enabling technology for the exploitation of new as well as existing materials, and for the development of novel design concepts and structural configurations. This potential has been helped by developments in adhesive materials and simplified surface treatment processes, together with an improved understanding of bonded joint behaviour and design concepts.

Adhesives can offer real technical and economic benefits, providing greater freedom and flexibility for designers. The benefits available may stem particularly from repair and strengthening work, as well as from opportunities with lightweight structures or those in which increased fatigue resistance is desirable. Largely to circumvent some of the potential on-site fabrication difficulties, a number of future development possibilities lie with the off-site prefabrication of bonded assemblies. A number of examples exist currently in building construction, such as sandwich panels, curtain walling and cladding assemblies, structural glazing, stressed skin flooring and framing, and 'macro' composite and compound structural elements. It is likely that greater use will be made of high strength alloys and polymer composite materials in the future.

The potential opportunities for making bonded connections in civil engineering structures outlined in this chapter are by no means exhaustive. They represent examples which vary substantially in developmental progress from possibilities to well researched concepts.

8.2 Sandwich construction

Sandwich construction describes an assembly of flanges and webs of great structural efficiency in bending; a lightweight low-modulus shear-resisting core or 'web' is implied sandwiched between two outer high-modulus skins or 'flanges'.

Open sandwich construction

This is the name given to the composite bridge deck slabs developed by the Wolfson Bridge Research Unit at Dundee University (1–3). The concept was developed with the object of developing a bridge deck with a weight saving over the conventional alternatives. Because a large part of the cost of medium and long span bridges is devoted to supporting the weight of the roadway deck, a lighter deck should reduce the cost of the supporting structure. A lightweight precast decking element system was therefore developed to span between the top flanges of longitudinal stringers and cross-beams. In principle, the concept may be likened to a more efficient form of the dished steel buckle-plates supporting a mass concrete base for roadway surfacing, a form of roadway deck that was employed on many steel bridges built between 1890 and 1935.

The materials chosen to comprise the skin and core were steel and concrete, respectively. A top plate or skin was found to be more or less redundant in quasi-static bending and shear, giving rise to the term 'open sandwich'. The concept based upon a flat soffit plate is illustrated schematically in Fig. 8.1. The combined thickness of steel and concrete is of the order of 150 mm, and the resulting precast unit is about 30% lighter in weight than a conventionally reinforced concrete slab of similar strength because (a) there is no concrete cover below the plate, and (b) the plate carries biaxial stress which, together with a reduction in dead weight from the lack

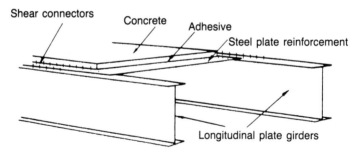

Fig. 8.1. Open sandwich construction.

of cover, enables a reduction to be made in the depth of concrete above the plate. During fabrication the steel plate acts also as the soffit formwork, thereby enhancing any potential economic benefit. In a typical medium-span bridge, the lighter deck can lead to a reduction of 10–15% in the cost of the supporting structure.

These 'open sandwich' slabs may be based on either a flat soffit plate or a plate curved slightly upwards to form a shallow arch, or inverted catenary. Their applications extend to repair and mainten-ance work, as well as in new construction. For instance, interest in this concept has been expressed in America for the replacement of existing highway bridge decks suffering from rebar corrosion. Upgrading through enhanced structural performance may be con-ferred additionally by the formation of a replacement composite deck with overall continuity. These slabs may also have application as a flooring medium in building structures, perhaps in competition with the use of cast-in-place decks utilising profiled steel sheeting. Alternatively, slabs could be prefabricated using profiled steel plates as the bottom plate or skin, with the adhesive providing the shear key. A number of possibilities exist, including the substitution of the main steel and concrete component materials for alternatives.

Precast open sandwich panels for bridge decks

Structural design. The comments which follow are applicable to the case of open sandwich slab elements of either constant depth or inverted catenary section acting compositely with steel plate girders, in accordance with BS 5400 (4). For constant depth elements an initial design approach may be to assume that the deck behaves as a series of laterally connected T-beams. The grillage method may then be used for an analysis of the global load distribution, although a number of caveats should be observed(5, 6). Analysis of the inverted catenary slabs is probably best accomplished using the finite prism technique(7), together with some local section design based upon conventional reinforced concrete theory. For initial sizing of the precast units, reference may be made to Table 8.1 which provides trial concrete and steel thicknesses for a range of longitudinal beam spacings.

Conventional bar reinforcement is advised for incorporation in the top surface of open sandwich decks to control cracking and to provide continuity. Tensile stresses will occur in the transverse

Table 8.1. *Initial sizing of open sandwich panels*

(a) *Constant depth units*

Plate girder spacing (mm)	Overall slab depth (mm)	Soffit plate thickness (mm)
1800	120	3
2300	130	3
2800	140	4
3500	160	4

(b) *Inverted catenary units*

| Plate girder spacing (mm) | Slab depth(mm) | | Soffit plate thickness (mm) |
	At haunch	At crown	
1800	200	50	3
2300	215	55	3
2800	250	70	4
3500	280	80	4

direction over the main beams in all decks and longitudinally over intermediate supports of continuous decks. Some suggested details illustrating the reinforcement suggested to accommodate these local stresses are incorporated in Figs. 8.3–8.6.

Punching shear failure is to be guarded against in both types of deck, and particularly over the relatively thin crown of inverted catenary sections. It seems as though the guidelines in CP110 (8) will provide adequate resistance in this respect, based upon laboratory studies(9). Fatigue behaviour of such slabs has been demonstrated to be very good, both in laboratory investigations and under road-traffic conditions.

Manufacture. Slabs consist of a thin steel plate with adhesive against which concrete is cast. Plate thicknesses should always be at least 3 mm, and the plate's surface(s) should be cleaned, degreased and gritblasted prior to bonding. Adhesive requirements are broadly similar to those specified for repair and strengthening purposes (see Chapter 6), although particular regard should be paid towards the amount of free moisture associated with fresh concrete and the highly alkaline nature of hardened concrete as an adherend.

A flexibilised cold-curing epoxide adhesive containing an aromatic amine hardener is likely to be the most successful, particularly if used in two layers or in conjunction with a priming layer. Alternatively an adhesive containing a modified aliphatic polyamine hardener may be suitable, in conjunction with a primer recommended by the manufacturer. The adhesive layer should be about 1 mm thick, and the concrete cast against it whilst this is still tacky. A rapid curing adhesive is not necessary; one which cures to the required mechanical properties in about 7 days will suffice. However, in view of the likely sustained loading to be borne by the adhesive, its flexural modulus and T_g should not be too low in order to avoid creep effects. The underside of the plate should be painted, or at least primed, as part of the final overall corrosion protection system.

Conventional or lightweight aggregate may be used to construct the slabs. The fresh concrete should be applied to the adhesive coated plate within 40 minutes of mixing the adhesive, requiring a carefully monitored working sequence. Compaction should be effected using either internal or external vibrators, and curing should be conducted in the normal manner. Curing at elevated temperatures, particularly using steam, is *not* recommended.

Installation. The installation of slabs into a bridge deck requires the careful consideration of a number of details. Firstly, lifting eyes must be incorporated into the tops of such slabs which are connected to the soffit steel. Secondly, the method of ensuring composite action and continuity must be decided upon. Composite action with the longitudinal beams may be assured by the use of shear studs, welded to the top flanges of the beams, which project through pre-drilled holes along the projecting soffit steel of the slabs. The slabs may also be 'bedded' on a layer of adhesive applied along the flange tops of the longitudinal beams. This provides shear resistance as well as maintaining a good seal between the slab and the beam. *In situ* concreting between the slabs can then be carried out after continuity reinforcement has been placed.

The continuity details outlined here assume that the open sandwich deck has been designed for full transverse continuity over the longitudinal plate girders, although a simpler detail may result if the slabs were assumed to be simply supported between the plate girders. The suggestions here are based partly upon experience gained during laboratory trials, but mainly on those employed in the construction of a 6 m span bridge constructed for Tayside Regional Council's Roads Department to the south of Loch Tummel (Fig. 8.2)(10). The structure has been in service since 1985.

271

Fig. 8.2. Installation of open sandwich deck slabs on bridge south of Loch Tummel, Tayside.

Transverse continuity may be provided by using an *in situ* concrete connection containing looped reinforcement cast into the slabs (Fig. 8.3); an alternative arrangement using folded steel plates may also be used (Fig. 8.4). The primary shear connection is provided by shear studs welded alternately to the soffit plate and the longitudinal beam flange. It is important that the precast/*in situ* construction joint occurs well away from the beam flange to avoid excessive rotations and subsequent slab distress under traffic loads.

Longitudinal continuity between slabs in regions of sagging moments may be made by using a bonded or welded cover plate (Fig. 8.5). Top bar reinforcement is optional in this case, to control differential shrinkage between the precast and *in situ* concrete. In regions of longitudinal hogging moment, the top of the slabs will need to be reinforced to control flexural tension cracking. Conventional bar reinforcement may be cast into the slabs as shown by a typical joint detail in Fig. 8.6. Naturally it is important to provide a suitable corrosion protection method for the underside of the slab soffit steel. Normally, painting of the bridge soffit would be carried out during construction, and then within a normal maintenance schedule.

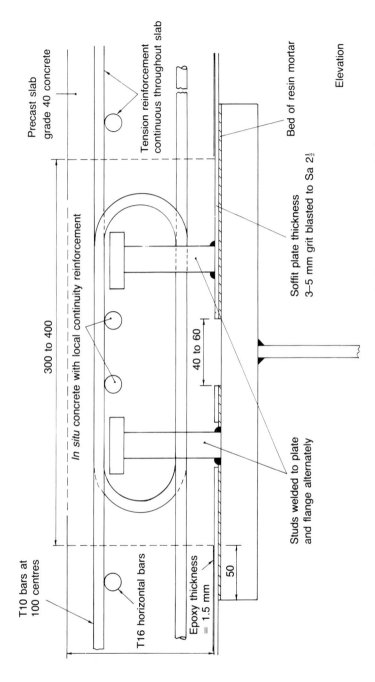

Fig. 8.3. Typical transverse continuity over main girder (high yield looped rebar).

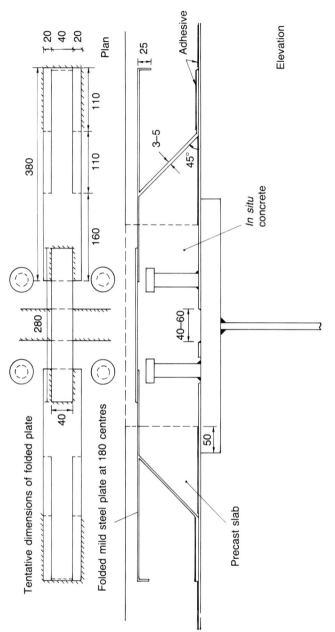

Fig. 8.4. Typical transverse continuity over main girder (folded mild steel plate).

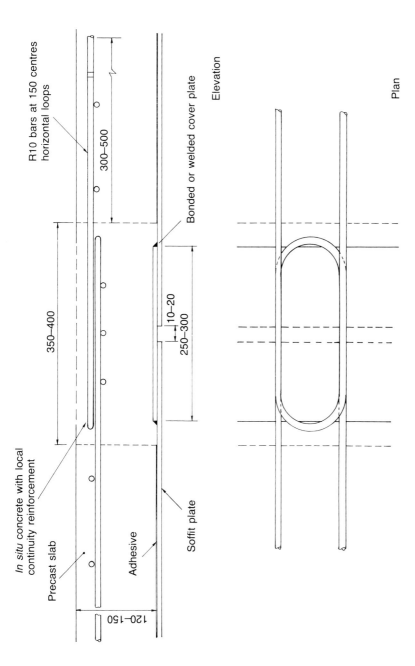

R10 bars at 150 centres
horizontal loops

300–500

Bonded or welded cover plate

Elevation

350–400

10–20

250–300

In situ concrete with local
continuity reinforcement

Precast slab

Adhesive

Soffit plate

120–150

Plan

Fig. 8.5. Typical longitudinal continuity joint detail in sagging region.

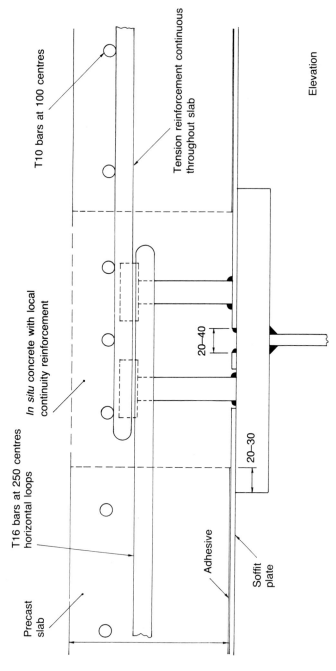

T10 bars at 100 centres

Tension reinforcement continuous throughout slab

In situ concrete with local continuity reinforcement

T16 bars at 250 centres horizontal loops

20–40

20–30

Precast slab

Adhesive

Soffit plate

Elevation

Fig. 8.6. Typical longitudinal continuity joint detail over an intermediate support.

Comparative example. A comparative costing exercise was conducted on the re-design of a typical continuous highway bridge with an open sandwich roadway deck (Fig. 8.7). The Tummel Bridge on the Pitlochry by-pass (A9), Tayside, consists of four steel plate girders which act compositely with a 225 mm thick reinforced concrete deck continuously supported over spans of 40 m, 70 m, and 40 m respectively. The study suggested that deck slabs consisting of a 150 mm thick concrete core and a 3 mm thick steel soffit plate would provide a suitable alternative deck slab. A re-costing established a saving of some 12% in superstructure cost if this form of deck was incorporated in designs carried out to latest Department of Transport specifications.

Closed sandwich construction

The possibilities for making bonded structural sandwich elements in a variety of materials are very real. However, whilst there exist structural examples such as aluminium honeycomb panels (used in aircraft and transport applications) and metal skinned foam sandwich panels (used as the monocoque chassis in refrigerated transport applications), these composite constructions are normally utilised in non- or semi-structural ways. Typical skin materials are steel, aluminium, GRP and plywood, and common core materials are rigid foam polystyrene, polyurethane, polyisocyanurate, PVC, and honeycombed aluminium. In some instances the foam core is injected between the skins and adheres to them; in others, adhesives are used to bond the separate components together. The nature of the manufacturing process depends on the type of structure to be made, and the degree of investment in production machinery. Both flat and complex curved forms can be made by a hand lay-up process as well as in an automated way.

The faces of skins or sandwich panels may be smooth, lightly profiled or heavily profiled, depending on the skin material and intended use. Polymer coated steels or aluminium are common materials for building components, and these may be readily profiled by cold-rolling to confer bending strength. The use of at least one profiled skin is essential for roof members which may be required to withstand long periods of static loading. The choice of colour, construction materials, overall dimensions and the fixing arrange-

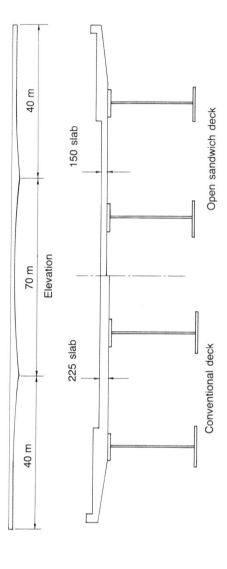

Fig. 8.7. Conventional and open sandwich bridge deck slabs.

ments remain important considerations because they can all influence the performance and long-term integrity of such panels.

Opportunities in civil and structural engineering arise in lightweight structures, or where the high strength-to-weight ratios can be exploited. These may be in bridging large spans or in the development of decking structures. The development of cellular decking structures employing steel skins and reinforced plastic I-beams has been proposed(11), for which full composite action and maximum bending efficiency could be achieved by bonding the flanges of the I-beams to the skin.

8.3 Composite construction

Steel/concrete

Adhesives may be used to bond precast concrete slab units directly to the flanges of longitudinal steel girders in bridge construction. Such a technique has been successfully employed in Germany(12). On one bridge described in Chapter 7, epoxy resins were used in a variety of forms and locations on the superstructure. The successful use of resins in this example suggests considerable scope for other applications.

Other material combinations

There exists great potential for the development of 'macro composite' assemblies for use as structural elements or members, in which adhesives could provide the stress transfer mechanism between the different constituent materials. Examples might include decking elements or roof members comprising steel and timber, for example. An interesting structural possibility has been investigated by Oxford Polytechnic within the ABCON Programme(13), in which a composite flooring assembly was developed by bonding particle board sheeting over the top flanges of cold-formed galvanised steel joists (Fig. 8.8). Considerable increases in bending stiffness and overall strength were realised when compared with the alternative use of conventional self-drilling, self-tapping, screws – even when employed at close spacings. Apart from the immediate benefit of enhanced stiffness and feeling of solidity, the development suggests

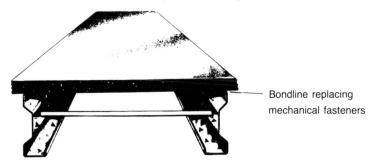

Bondline replacing
mechanical fasteners

Fig. 8.8. Bonded steel joist/particleboard flooring connections.

either that economies can be made (in particle board thickness or joist material) or else that much stronger, stiffer assemblies can be fabricated.

The concept of bonding skins to frames, as opposed to the use of discrete mechanical fasteners, represents an obvious application of adhesives in order to stiffen existing structural elements by introducing stressed skin or membrane action. This is really an extension of the composite flooring development described above which examined the bending performance of assemblies constructed with long bondlines. It is also analogous to the use of polyurethane adhesives for the direct glazing of car bodies which confers significant additional body stiffness (see Chapter 1). A number of possibilities exist in the construction industry, perhaps the most obvious being the bonding of wall panels to lightweight structural steel framing systems, and the bonding of cladding panels to curtain walling systems.

8.4   Compound structural elements

Useful mono-material structural elements may be formed by bonding together a number of individual pieces of the same material. Glued laminated timber (glulam) members as described in Chapter 7 represent a development in which structural elements of large cross-section may be created. Adhesive bonding, as opposed to mechanical fastening, leads to reduced stress concentrations as well as the elimination of joint slip. The same principle may be applied to other materials where appropriate, such as aluminium extrusions and polymeric pultrusions. The high strength-to-weight advantage than can be realised would normally be the motivating factor.

Aluminium extrusions

There is a technical and economic limit to the size and complexity of shape of a single aluminium extrusion. This limits their structural possibilities unless groups of extrusions can be joined together to produce large or intricate multi-chamber hollow combinations. By joining such groups together using adhesive bonding, as opposed to welding, there is greater choice of alloy type. This concept of combining individual extrusions has been advanced by British Alcan for transport and construction possibilities. The deep I-beam development aimed at the offshore industry measures some 12 m long by 1 m deep, and comprises fifteen individual extrusions bonded with a single part heat-cured toughened epoxide (Fig. 8.9). Elements such as these may have significant potential for exploitation in, say, wide-span column-free building structures. In the Alcan developments the tapered mating surfaces of the joints were gritblasted and primed with a silane coupling agent prior to bonding.

Composite pultrusions

As with aluminium extrusions, so with pultruded fibre reinforced composite profiles. The limit to the size and complexity of these profiles suggests that a modular approach could be adopted towards forming alternative structural configurations from the basic or standard profile shapes by bonding together individual lengths. Composite materials lend themselves to being joined with resin adhesives because they are themselves formed with vinyl ester, polyester or epoxy resins. Cursory surface treatments only, such as mild abrasion, often suffice.

8.5   Steelwork fabrication

The potential for adhesive bonding as a substitute for welding (or bolting) of steel structures is very large. In the short term, bonding

281

Fig. 8.9. Compound bonded aluminium I-beam comprising 15 individual extrusions (courtesy British Alcan Aluminium).

techniques may prove to be particularly useful when an existing structure requires to be either stiffened or strengthened. Some advantages include:

(1) an improved fatigue detail
(2) lack of distortion from, say, the heat of welding
(3) no loss of metal section, which is inevitable with bolting.

A resulting improved fatigue detail arises through the avoidance of residual stresses, which can dominate the fatigue life of relatively thick section assemblies. In thinner section structures the consequence of these stresses is visible distortion, often requiring expensive finishing operations.

Bonded stiffeners

Plate structures which have been stiffened by the addition of welded attachments can suffer from the problems of distortion and inferior fatigue performance outlined. The use of adhesive to unite the stiffener and the plate has been the subject of considerable research interest, and shows much promise. Harvey and Vardy(14) reported on an investigation into the performance of bonded shear stiffeners on plate girders (Fig. 8.10), whilst Hashim *et al.*(15) have examined the potential for using bonded stiffened plates in ship hulls. In the former research work a number of half-scale girders were built and tested with over- and under-sized bonded T-stiffeners (Fig. 8.11); the failure loads were somewhat higher than those predicted by finite element modelling. In addition to the buckling tests performed on complete girders, a number of fatigue tests were also carried out on specimens mimicking the web/stiffener cross-section. This work was very encouraging, and a number of further opportunities would seem to exist with the addition of stiffeners to existing structures.

With bonded joints involving relatively thick stiff steel sections the strength and performance of such joints is governed largely by the adhesive's properties. The need for rigidity and performance at higher temperatures than with concrete structures dictates the use of stiff high performance adhesives. So far, single-part heat-cured toughened epoxies have been identified as suitable products. Gritblasting of the adherend prior to bonding, perhaps with the addition of a silane primer, should ensure satisfactory long-term performance.

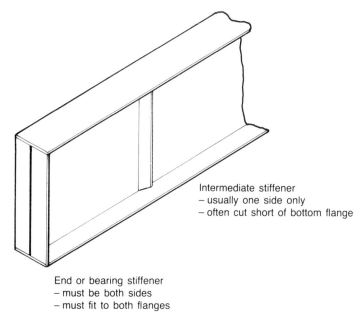

Intermediate stiffener
– usually one side only
– often cut short of bottom flange

End or bearing stiffener
– must be both sides
– must fit to both flanges

Fig. 8.10. Plate girder-bonded intermediate web stiffeners.

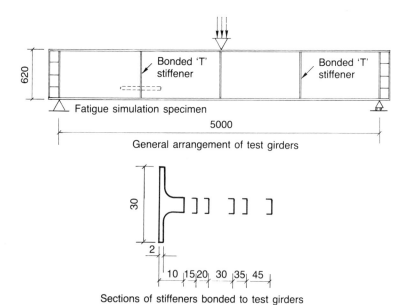

620

Bonded 'T'
stiffener

Bonded 'T'
stiffener

Fatigue simulation specimen

5000

General arrangement of test girders

30

2

10 15 20   30 35 45

Sections of stiffeners bonded to test girders

Fig. 8.11. General arrangement of test girders with bonded web stiffeners (Ref. 14).

Tension splices, beam splices and cover plates

The feasibility of using adhesives, in combination with bolts, to bond common structural details on highway bridges has been examined carefully by Albrecht *et al.*(16–18). The motives behind their work were to (a) increase the fatigue strength of steel bridge members, (b) reduce the connection size without lowering the fatigue strength, and (c) eliminate welded details with low fatigue strength. A structural acrylic adhesive was selected for the tests described below(17); surface treatment involved shot-blasting.

A number of tension splices using various combinations of bolting, fabricated with and without adhesive bonding, were fatigue loaded (Fig. 8.12). The results showed that bonding, in addition to bolting, consistently increased the fatigue life of the joints, offering the potential to reduce the number of bolts to develop a specified mean fatigue strength.

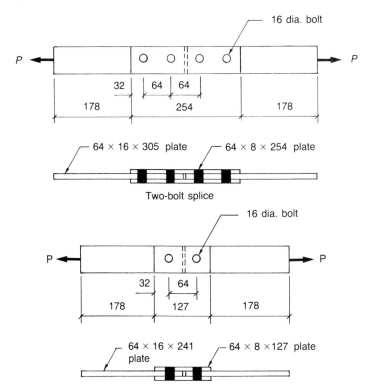

Fig. 8.12. One- and two-bolt tension splices (Ref. 17).

285

Several beam specimens were fabricated consisting of rolled beams whose flanges and webs were spliced at mid-span (Fig. 8.13). The beams were tested in four-point bending, with the splice located in the region of constant bending moment. The flanges were spliced with plates of different sizes, and carried the moment. The undersized web splices were provided for ease of specimen fabrication and carried no shear. Bolting, with and without bonding, was employed. Again the results showed that bonding, in addition to bolting, increases the fatigue life of the detail, offering the potential to reduce the number of bolts in order to develop a specified mean strength.

Some rolled beams were additionally fabricated with cover plates bonded to the tension flange (Fig. 8.14). The original intention was to use adhesives only to join the cover plate to the flange, but preliminary fatigue studies indicated debonding of the plate-ends. High strength bolts were therefore introduced to negate the tendency towards peeling. When this was done, the fatigue life was increased by a factor of 20 over that of conventionally welded cover plates.

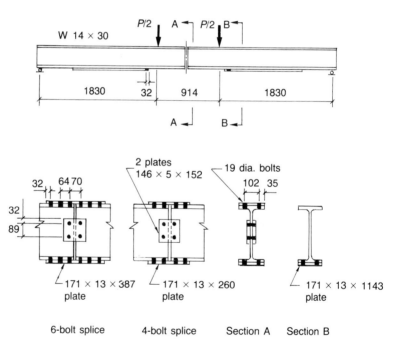

Fig. 8.13. Four- and six-bolt beam splices (Ref. 17).

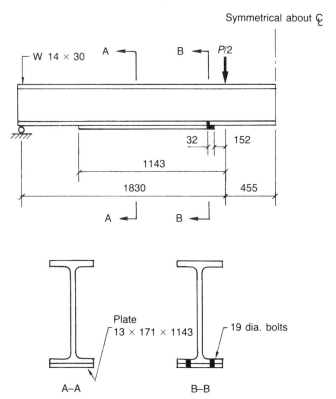

Fig. 8.14. Cover-plated beam specimen (Ref. 18).

The study concluded that the largest increase in the fatigue strength of structural steel details was obtained by replacing welded connections with bonded and bolted connections. When properly detailed, such connections would be virtually fatigue-proof for highway bridge loading. However, it would be desirable to find adhesives with superior creep resistance and durability to cope with the expected service life.

Hollow sections and truss connections

Joints between, and attachments to, rolled hollow sections would seem to provide opportunities for adhesive bonding. For instance, the attachment of handrails to pedestrian overbridges constructed from hollow sections such as vierendeel girders, or the fabrication of gusseted or socketed nodes and joints between members. The use of adhesives would also provide an effective seal to the hollow sections. Similarly, adhesives could be used to fabricate suitably designed joints in trusses comprising various structural sections.

8.6   Connections between metals

Light-gauge materials

Adhesive joints between relatively thin gauge metals are common throughout the engineering industry. Such metals are used extensively in steel and aluminium curtain walling assemblies, and in proprietory steel framing systems. Cold-formed steel sections are also used extensively in building construction. The components of the walling and framing assemblies are commonly welded or brazed, but this can be detrimental; where coated metals are involved, it can be very difficult or impossible. There is an increasing trend towards the use of inorganically and organically coated metal products whose protection is lost during joining operations other than by adhesive bonding. The scope for the increasing use of adhesives in such applications is therefore very great.

Aluminium connections

The bonding of aluminium alloy components for structural engineering applications has been the subject of extensive research by the Dutch TNO Institute for Building Materials and Structures(19, 20). Apart from the evaluation and testing of a number of adhesive systems, experimental research was carried out on several structural details. Aluminium alloy surface pretreatment was by degreasing only, to represent a 'practical' procedure.

Three-point bend tests on beams of hollow rectangular section, fabricated from angle sections and flat plates bonded together, demonstrated the importance of adhesive ductility and toughness; beams fabricated with a relatively brittle epoxide split apart at the onset of buckling, whereas those bonded with tougher adhesives behaved in a more ductile manner.

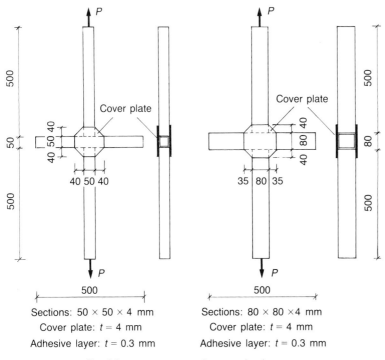

Sections: 50 × 50 × 4 mm
Cover plate: *t* = 4 mm
Adhesive layer: *t* = 0.3 mm

Sections: 80 × 80 ×4 mm
Cover plate: *t* = 4 mm
Adhesive layer: *t* = 0.3 mm

Aluminium pretreatment: degreased only

General test arrangement

Fig. 8.15. Bonded aluminium cruciform joints (Ref. 19).

A number of cruciform joints of rectangular hollow section were fabricated with adhesive gusset plates (Fig. 8.15). Hollow sections of two different sizes were used, and some of the specimens were subjected to accelerated ageing prior to testing in tension. The structural behaviour of these joints was also predicted by non-linear, three-dimensional, finite-element analysis. Very good agreement was obtained between the experimental and predicted load–displacement behaviour.

The performance of curtain walling wall panels fabricated with bonded and with bolted channel section stiffeners was compared. Under simulated wind pressure, the bonded stiffeners accommodated gross panel deformations representing several times the design requirements.

289

Some beam specimens were fabricated whose flanges and webs were spliced at mid-span in a similar way to Fig. 8.13. The performance in four-point bending was encouraging, although yield in the aluminium alloy was rarely experienced.

Several semi-structural elements with polyurethane bonded cover plates were also tested, demonstrating ample performance characteristics.

In general, two-part cold-cured modified epoxides and modified acrylics were found to meet the strength and deformation requirements for structural bonding. However, the strength levels attained with joints made with the acrylics were somewhat lower. Longer-term performance and durability remain subjects for further evaluation.

Rebar connectors

The splicing of reinforcement bars using sleeved couplers offers a number of advantages. These include the avoidance of rebar congestion, positive stress transfer, reduced crack widths caused by bursting forces at lapped joints, and the ability to replace or add reinforcement in concrete repair work. Sleeve joints relying on epoxy resin only were investigated by Navaratnarajah(21), but more recently Ancon (MBT) Couplers(22) have marketed a mechanical locking coupler whose fatigue performance may be enhanced by the injection of an epoxy resin. The coupler consists of a steel barrel in which there are various locking bolts and internally machined load transfer grooves. The positive mechanical key provides the necessary resistance to fire and creep which the resin cannot offer.

8.7   Connections between plastics

About 30% of all polymers produced each year are used in the civil engineering and building industries(23). Nevertheless, structural plastics such as fibre reinforced composites have so far received little attention by civil and structural engineers, despite some of their obvious advantages such as lightness, handleability and corrosion resistance. This may be due to reservations on credibility grounds or fire resistance properties, as well as to uncertainty on how to design structures with them. Whilst their mechanical properties are in fact fairly well understood, there are a number of problems concerned with making joints with and between these materials. Currently mechanical fasteners such as bolts are utilised but adhesives, either alone or in addition, can offer significant advantages.

In building and construction, pipes probably represent the largest volume of plastics usage whilst engineered plastics have found their way into structural frames, platforms and walkways, and for use as post-tensioning tendons on some concrete bridges in Germany (at Dusseldorf and Berlin)(23, 24). For very aggressive environments, pultruded fibreglass reinforcing bars are beginning to be accepted in concrete constructions. In the future, bundles of pultruded fibres may be formed into lighter weight cables for suspension bridges, to enable very long spans to be accommodated. At a smaller scale it may be possible to exploit filament wound hollow sections in trusses and space frame structures. Current research effort is being directed towards the problems of joining, by bonding, tubular composite members for use in plane and space grid frames(23).

Pultruded sections

Glass reinforced pultruded plastic sections such as Extren(25) have been used where corrosion resistance, thermal and electrical non-conductance or non-magnetic properties were required. It is possible to fabricate large assemblies using bolting techniques (e.g. Fig. 8.16), but development work is necessary to explore the opportunities for bonding.

Cellular panels, manufactured by GEC Reinforced Plastics (Fig. 8.17), were installed in 1988 as a permanent access decking and enclosure medium to the underside of the A19 Tees viaduct near Middlesborough. 16 000 m² of panelling were hung from the steel plate girders, each panel measuring some 12 m long by 3 m wide(26). Although adhesives were not employed directly in this application, the concept of the double skinned panels and their design indicates the possibilities for using adhesives in joint configurations which incorporate mechanical interlocking.

Cladding attachments

There exist a number of instances in which cladding panels, comprising various materials, are bonded directly into curtain

Fig. 8.16. Communications equipment turret constructed of EXTREN on top of the Sun Bank building in Orlando, Florida.

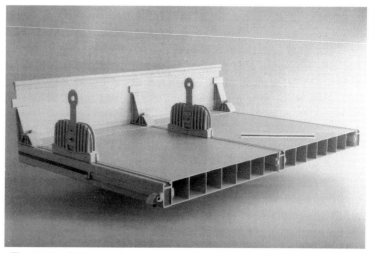

Fig. 8.17. GRP cellular panel employed for Tees viaduct enclosure.

walling assemblies using toughened acrylics or structural silicones – predominantly under factory conditions rather than on site. In some cases the design may accommodate a degree of additional mechanical restraint, particularly if required by local building regulations. Nevertheless there remain several opportunities for using adhesives in cladding, particularly with heavily filled polyester sheet panel materials. Adhesives can be used for 'secret fix' applications to bond stiffeners to the reverse of panels, or to bond fixing attachments to them, or to bond them directly into frames. Apart from eliminating visible external fixings and problems of 'read through' on thin or opaque panels, the assembly process is made simpler and more economic.

Externally bonded reinforcement

The technique of bonding steel plate reinforcement to the tension and shear faces of existing reinforced concrete structures for strengthening purposes was described in Chapter 6. As an alternative to steel plate reinforcement, glass or carbon fibre reinforcement could be used which may (a) simplify some of the surface treatment procedures, (b) reduce installation costs because of the ease of handling and joining such materials, and (c) eliminate the costs associated with subsequent corrosion protection and maintenance of steel plates. Glass, carbon or mixtures of both fibres could be used and some preliminary evaluation work in the US has been reported(27). Development work has already been conducted on the selective stiffening and strengthening, or repair, of lightweight portable aluminium Bailey bridges used by the military. Strips and plates of fibre reinforced plastic were bonded to the surfaces where necessary using cold-curing adhesives.

Plastic bridges

Short span glass reinforced plastic (GRP) bridges have existed for some time in Europe and in the USA, but have carried only pedestrian traffic. In 1982, a 20 m span highway bridge was completed in Beijing, China(28). It comprises five honeycomb core box girders, each 21 m long, 1.6 m wide and 1.7 m deep, fabricated by the hand lay-up of layers of glassfibre cloth and polyester resin. The girders were then united to make a composite deck with further layers of glassfibre cloth and resin. The exposed surfaces of the bridge are coated with a resin coating to resist UV degradation. The largely experimental structure weighs only one fifth of that of an equivalent reinforced concrete counterpart, and demonstrates the potential savings on a longer span structure using GRP elements.

8.8 Miscellaneous connections

There exist a number of very diverse uses for adhesives in both new work and in existing structures. Nevertheless, in such cases the adhesive would generally be called upon to act in a structural manner.

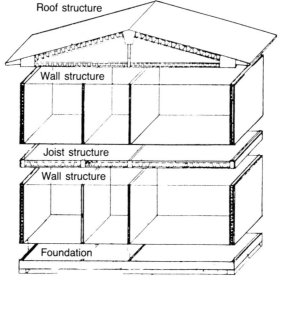

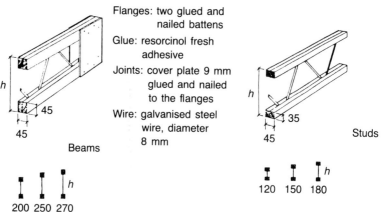

Fig. 8.18. Wirewood.

294

A similar concept to glulam and concrete segmental construction can involve the fabrication by bonding, usually with epoxides, of awkward or special shapes and units made of concrete or clay brick. Examples include 'special' roof tiles and building blocks formed by cutting and bonding standard production shapes to provide new forms.

Wirewood describes a Swedish structural innovation in beams and stud partitioning, and consists of timber flanges connected by steel bars which are mechanically restrained and bonded into position (Fig. 8.18). The assemblies can be formed into various sizes and depths, and be used as floor joists, in wall frames, or in roof trusses (Fig. 8.19). A simple cost-effective modular approach to building is claimed by European Building Components who market the system. Another use of adhesives in bonding steel to wood has involved the incorporation of flitch plates for making hidden connections between glulam members(29). The adjacent members of a portal frame sports centre structure in Dundee were slotted, in section, to accommodate a steel plate which was both epoxy-bonded and bolted through the timber. Load testing indicated that the configuration behaved as a composite section over the length of the connection. Widespread use is also made of bonding steel bolts into glulam structures in Scandinavia, further demonstrating the confidence in such steel/wood bonded connections.

Resin anchors are quite widely used already for fixing steel bars and ties in concrete and masonry structures. A concept marketed

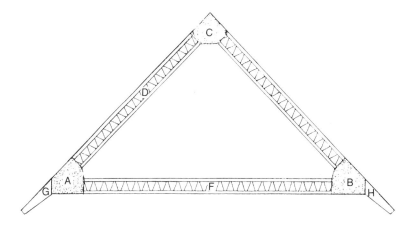

Fig. 8.19. Truss assembled from Wirewood.

more recently by Pitchmastic(30) enables the retro-fixing of wall ties, structural support dowels and inclined panel ties into cavity walls or poor quality masonry (see also Fig. 6.26). A polyester resin is injected into the annulus formed by drilling out the substrates, and a polypropylene sleeve is used to bridge the cavity, if required. The remedial wall tie market is growing steadily, and versatile resinous solutions will be required increasingly.

8.9   Closing remarks

Adhesive materials will continue to have a significant part to play in the construction industries, and the contribution of structural adhesives can only expand in the future. Additionally, engineers will need to develop some understanding and expertise in the field of adhesion science and technology if greater advantage is to be taken of the potential waiting to be exploited.

Many opportunities exist currently in the remedial markets, in the fabrication of aluminium and steel assemblies and, in particular, for stiffened plate structures. Great advantages can be derived from combining material combinations and, indeed, from combining adhesive bonding with some form of mechanical attachment to create composite harmony.

In the future it would seem that greater use will be made of materials combinations and of advanced composite materials – perhaps associated with space exploration. Lighter-weight structures will be increasingly demanded for many applications and bonded stressed-skin concepts are likely to find increasing favour. Adhesives will have an invaluable role to play in all such instances.

The adhesives themselves will continue to evolve, particularly with a closer association of formulators and engineers, and it is to be expected that more surface-tolerant and durable systems will become available. The selection of suitable adhesives and the design of structural bonded joints will also be facilitated with the greater accumulation of knowledge and expertise by engineers. As the experience in many other industries suggests, the more widespread application of adhesives to bonding civil engineering structures remains only a matter of time.

APPENDIX

Compliance spectrum for a two-part cold-cure adhesive for structural bonding of steel to concrete

The sample specification, or 'compliance spectrum' which follows was introduced by the authors in 1988 and published in the Proceedings of the Institution of Civil Engineers(1). It addresses the requirements of adhesives, bonding procedures and test methods for structural steel-to-concrete bonding. The test methods recommended were based on the experience gained from extensive programmes of research into the control, classification and durability of structural adhesives for application in civil engineering undertaken at Dundee University.

A.1 Purpose

The purpose of the above is to specify requirements for a cold-cure adhesive to permit *either* the repair or strengthening of existing concrete structures by bonding on additional external steel plate reinforcement, *or* the construction of steel/concrete composite units in which wet concrete is poured on to steel freshly coated with a layer of adhesive. In both cases the adhesive serves to resist the interfacial shear stresses necessary to ensure structural composite action between the steel and concrete. For these purposes a cold-cure adhesive is defined as one which is capable of curing to the required strength between the temperatures of 10 °C and 30 °C.

A.2 Form of material

The adhesive should be a two-part epoxy comprising resin and hardener components. The resin will normally be based on the diglycidyl ether of 'bisphenol A' or 'bisphenol F' or a blend of the two. The hardener, or curing agent, will normally be from the polyamine group, since these tend to result in adhesives with better resistance to moisture than have the polyamides and are likely to cause less concern over creep performance under sustained load than the polysulphides. Other additives, such as coupling agents, diluents, flexibilizers, plasticizers, toughening agents, surfactants and inert fillers, may also be incorporated into the formulation to improve the application or performance characteristics of the adhesive. In the case of inert fillers, these may be supplied as an alternative third component for inclusion at the time of mixing.

If a filler is used it should be a non-conductive material and may be treated with a coupling agent appropriate to both filler and resin type. The filler should be highly moisture resistant, should withstand temperatures up to 120 °C without degradation, and should have a maximum particle size of 0.1 mm.

The toxicity of the chemicals used in both the adhesive and primer should be low enough to enable safe use in a normal workshop environment and on a construction site. They must satisfy the intent and requirements of the Health and Safety at Work Act, and if special ventilation is necessary the requirement should be specified in detail.

The adhesive should be supplied in liquid form and should mix readily to a smooth, paste-like consistency of initial viscosity between 20 and 150 Pa at a shear rate of 10^{-1} s, and have a yield stress of at least 20 Pa at 20 °C suitable for spreading both on vertical and horizontal surfaces(2). The resin and hardener should be of dissimilar colour to aid thorough mixing. The mixed material should be free of lumps, and the components, including filler, should not settle out or separate during the pot life of the adhesive.

Due regard should be paid to the practical difficulties of achieving a high standard of substrate surface pretreatment in repair and strengthening works on site. The stability of the adherend/adhesive interface is probably the most important factor in the durability of bonded joints.

Suitable primers of the corrosion inhibiting type should be specified for use with this adhesive, gritblasted steel adherends and hardened concrete.

Particular regard should be paid to the highly alkaline nature of concrete and any possible adverse effect on bond strength, especially in the long-term. For new construction, the amount of free moisture associated with pouring wet concrete on to the adhesive-coated steel should be noted. It is also likely that a joint configuration employing permeable adherends such as concrete will be inherently less durable than metal/metal joints.

Consideration should be given to the difference in elastic modulus and thermal characteristics of the steel, concrete and adhesive.

A.3   Working characteristics

Application. The adhesive should be capable of being applied readily to both concrete and treated steel surfaces in layers from 1 to 10 mm thick, to allow for concrete surface irregularities. Proprietory dispensing equipment is available to facilitate correct metering and mixing of the adhesive.

Usable life. The usable life of the mixed adhesive before application to the substrate should be not less than 40 minutes at 20 °C; that is, the viscosity should not rise above 150 Pa within this time.

Open time. The open time starts when the adhesive has been applied to the parts to be joined, and it represents the time limit during which the joint has to be closed. It should be not less than 20 minutes at a temperature up to 20 °C.

Storage life. The storage (or shelf) life of both the resin and hardener should be not less than six months in original containers at 5°–25 °C.

Curing time and temperature. The adhesive should be capable of curing to the required strength at temperatures between 10 °C and 30 °C in relative humidities up to 95%. For applications involving repair or strengthening, the adhesive should cure sufficiently to confer the specified mechanical properties at 20 °C in not more than three days. In new construction, this time period may be extended to seven days. The adhesive should undergo negligible shrinkage on cure.

A.4   Mechanical properties of hardened adhesive and steel-to-steel joints

A cure and storage temperature of 20 °C should be used for all tests described in this clause. Results should be based on the mean of a minimum of five tests in each case. Notwithstanding the effects of test rate and temperature, it should be recognized that the mechanical properties of the adhesive may undergo significant changes if plasticized by absorbed water.

Hardened adhesive

Moisture resistance. The adhesive should be formulated to minimise moisture transport through the adhesive itself. The equilibrium water content (M_∞) should not exceed 3% by weight after immersion in distilled water at 20 °C. The permeability, obtained from the product of the coefficient of diffusion (D) and M_∞, should not exceed 5×10^{-14} m²/s at 20 °C (see Fig. A.1). A film of adhesive, approximately 1 mm thick cast in polytetrafluoroethylene-lined moulds and weighing at least 3 g, is suggested for this test(3). This requirement is to enhance the potential for a durable adhesive/adherend interface, even if moisture uptake is not deleterious to the adhesive itself.

Temperature resistance. The adhesive should have a heat distortion temperature (HDT) of at least 40 °C. A specimen 200 mm × 25 mm × 12 mm deep tested in four-point bending, as shown in Fig. A.2, is recommended(4). The sample under test is placed in a temperature-controlled cabinet at 20 °C and a load

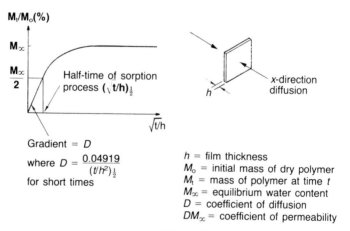

Fig. A1. Fickian diffusion.

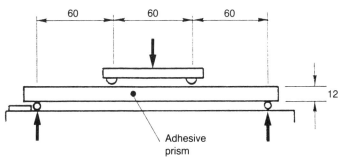

Fig. A2. 'All-adhesive' flexural test.

applied such that the specimen is subjected to a maximum fibre stress of 1.81 MN m^{-2} in accordance with BS 2782 (5). The HDT of the adhesive is taken as the temperature attained, as measured on a thermocouple attached to the specimen, after undergoing a further 0.25 mm deflexion while subject to a surface heating rate of 0.5 °C per min. This requirement is primarily to enhance creep resistance.

Flexural modulus. The instantaneous flexural modulus of the adhesive at 20 °C should be at least 2000 MN m^{-2}. A specimen 200 mm × 25 mm × 12 mm deep tested in four-point bending, as shown in Fig. A.2, is recommended(4). The sample under test is loaded at the third points at a rate of 1 mm/min and the central deflexion recorded. From the load–deflexion curve the secant modulus at 0.2% strain is calculated. This requirement is to assist in preventing problems due to creep of the adhesive under sustained loads. It should be noted that the higher the flexural modulus of the adhesive, the more susceptible the adhesive will be to stress concentrations arising from strain incompatibilities, for example at changes of section. For this reason, an upper limit of between 8000 and 10000 MN m^{-2} is probably desirable.

Shear strength. The bulk shear strength of the hardened adhesive at 20 °C should be at least 12 MN m^{-2}. A specimen 200 mm × 12 mm × 25 mm deep tested in a shear box, as shown in Fig. A.3, is recommended(4). The complete assembly is loaded in compression through a spherical bearing at a rate of 1 mm/min. The failure load is taken as that load at which shear cracking first

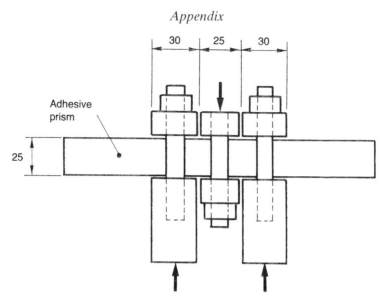

Fig. A3. 'All-adhesive' shear box test.

occurs. This requirement ensures that the adhesive will be at least as strong as the concrete to which it is to be bonded.

Tensile strength. The tensile strength of the hardened adhesive at 20 °C should be at least 12 MN m^{-2}. A dumb-bell specimen of dimensions according to BS 2782 : Part 3 (5) (methods 320 B/C) and having a cross-section of 10 mm × 3 mm is recommended (see Fig. A.4) (3). The specimens should be cast in polytetrafluoroethylene-lined moulds. Adhesive ductility may also be measured with appropriate strain monitoring equipment.

Steel-to-steel joints

The surface preparation of steel adherends for joint testing should be as follows.

(1) If necessary, remove heavy layers of rust by hand, or by mechanical abrasion with emery cloth, or by wire brushing, to give rust grades A or B as defined by Swedish Standard SIS 055900 (6).

(2) Either wash with detergent solution, rinse thoroughly with cold

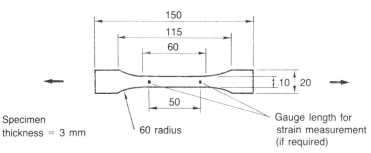

Fig. A4. 'All-adhesive' tensile test.

water and dry in a stream of warm air and with clean absorbent paper, or wipe with a suitable solvent and allow to evaporate.

(3) Gritblast to grade Sa2½ of Swedish Standard SIS 055900 (6) to achieve a mean maximum peak-to-valley depth of at least 50 μm, using a hard angular clean metal grit which is free of any grease contamination. For stainless steels, a non-ferrous grit – e.g. alumina based – should be used.

Abrasive dust remaining on the adherend surfaces after blasting should, ideally, be removed by brushing, blowing or vacuuming; solvent degreasing at this stage is not recommended. If a compatible primer is to be employed, all joint tests should be carried out using both unprimed and primed steel surfaces. Joints should be manufactured in jigs to provide a light clamping force during the cure, using carefully positioned internal wire or external steel spacers to control the adhesive thickness at a constant value between 0.5 and 1.0 mm.

Joint fracture toughness. The initial mode I fracture toughness K_{IC}, obtained from the mean of several tests using wedge cleavage specimens, an example of which is depicted in Fig. A.5, should not be less than 0.5 MN m$^{-3/2}$. The tests should be carried out at 20 °C using a wedge insertion rate of approximately 50 mm/min. The crack extension (a) should be measured on both sides of the specimen, e.g. by using a travelling microscope. The equation for the fracture of energy of the specimen is(7):

$$G_{IC} = \left(\frac{E_s d^2 h^3}{16} \right) \left(\frac{[3(a + 0.6h)^2 + h^2]}{[(a + 0.6h)^3 + ah^2]^2} \right)$$

303

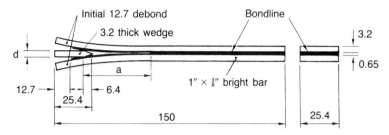

Fig. A5. Wedge cleavage specimen.

where E_s is the elastic modulus, and h the thickness, of the steel adherends. The fracture toughness K_{IC} can then be obtained by using the expression $K_I^2 = E_a G_{IC}$ for plane stress, where E_a is the elastic modulus of the adhesive obtained as described above. Such specimens may be further employed for stressed durability testing, by monitoring the rate and location of crack growth for different exposure environments(8).

Lap shear strength. Tests should be carried out over a range of temperatures, specifically including $-25\,°C$, $+20\,°C$ and $+45\,°C$, using bright mild steel adherends. The temperature should be measured by means of a thermocouple attached to the steel surface of the joint. Two alternative forms of lap shear joint may be employed.

(1) The thick adherend shear test (TAST) joint, as shown in Fig. A.6, is recommended for the determination of adhesive shear strength(3). The minimum lap shear strength required is 18 MN m^{-2} at 20 °C. The joint is fabricated from two bright mild steel strips which are bonded together and subsequently drilled and sawn as shown. The joint should be loaded to failure at a rate of 0.5 mm/min through universal joints. Adhesive spew along the sides of the short overlap region should be carefully removed before testing. This form of joint is preferred to those specified in BS 5350 : Part C5 (9), since a more uniform distribution of shear stress is obtained along the bondline and large peeling stresses are avoided.

(2) An alternative form of double overlap joint, which more closely reflects shear stress distributions in service, is shown in Fig. A.7. The dimensions l and d are selected so that the

304

Compliance spectrum

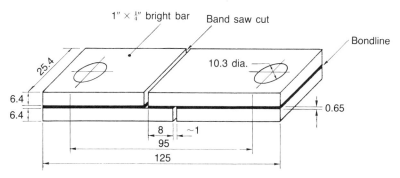

Fig. A6. Thick adherend shear test (TAST) specimen.

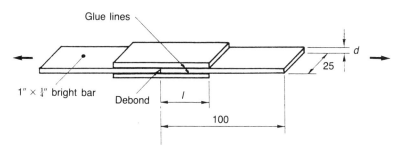

Figure A7. Double overlap joint.

bond ruptures before yield of the adherends(4). The minimum average lap shear stress required at failure is 8 MN m^{-2} at 20 °C. The ends of the main test pieces should be debonded to avoid load being transferred in tension, and any adhesive spew at the ends of the side laps should be carefully removed.

It is emphasized that the failure load of any adhesive joint is influenced by the joint geometry, rate of loading and test conditions. Therefore, lap shear tests should be used only for comparative purposes in adhesive assessment, selection and control.

Fatigue test. Fatigue tests should be carried out at a constant frequency between 1 and 25 Hz, with a sine wave form and using TAST or double overlap joints as described above. The joints should be capable of surviving, without failure, 10^6 cycles of a mean shear stress range between 1.0 and 10 MN m^{-2} at 20 °C for TAST joints

(Fig. A.6) (3) or between 0.4 and 4.0 MN m^{-2} at 20 °C for double overlap joints (Fig. A.7) (10).

Durability tests. The environmental conditions during the operation and the required length of service for civil engineering applications should be taken into careful consideration. As a strengthening technique, the minimum required life is thirty years while for new construction it may be considerably longer. Environmental conditions are likely to include a wide range of temperatures, as mentioned previously, combined with condensing humidities. In bridge or marine structures, the joint may also be subject to spray from de-icing salts or from the sea.

Accelerated laboratory tests employing, for example, the wedge cleavage test described above should be used to demonstrate, as far as is practically possible, that joints made with various combinations of surface pretreatments, primers and adhesives might reasonably be expected to satisfy these durability requirements(8). The collection of data from ageing in natural environments is most useful. For longer-term durability assessment, the TAST specimen, possibly in a stressed condition, could be employed. Adhesive dumb-bell specimens subjected to wet or moist environments may provide useful information on changes in strength and ductility through water-induced plasticization(3).

A.5   Quality control data

The manufacturer should keep on file the normal quality control test data for each batch of adhesive, and should provide a copy of the test reports for each batch on request. He should also certify conformance with this Specification. Average results should meet the Specification requirements and no single result may be more than 15% below specification.

A.6   Packaging

The adhesive should be packed in containers of suitable sizes so that weighing or measurement is not required. That is, the correct proportions should be obtained by mixing the entire contents of components A and B (and C).

Each container should be durably and legibly marked with

(1) manufacturer's name and address
(2) batch number and date of manufacture
(3) date of expiry or shelf life
(4) conditions of storage necessary to achieve the specified storage life
(5) mixing conditions
(6) safety precautions and details of any health hazards resulting from improper use.

A.7 Instruction sheet

The manufacturer should provide a dated, coded and titled instruction sheet with each delivery of adhesive. It should refer to this Specification and indicate conformance with it. It should also state

(1) the general chemical type of each component used in the adhesives
(2) recommended storage conditions and shelf life when stored under these conditions
(3) preparation instructions for steel and hardened concrete surfaces
(4) instructions for use of the primer, including optimum thickness and suggested range of permissible thickness
(5) mixing instructions, including allowable variations in mix ratio and any temperature control requirements during the mixing process
(6) application instructions, including limits on pressure, temperature, open time and relative humidity before mating the faying surfaces; these should also state whether adhesive should be applied to one or both adherends
(7) maximum allowable interval between application of primer to coated steel or concrete, and application of adhesive and any primer reactivation procedure if applicable
(8) safety precautions for all components of the adhesive and primer
(9) curing conditions, including the amount of pressure to be applied, the length of time under pressure and the temperature of the assembly when under pressure; it should be stated whether this temperature is that of the adhesive layer or of

the atmosphere to which the assembly is to be maintained or both. A graph of cure time against temperature should be supplied

(10) condition procedure before testing or use of the assembled product, including length of time, temperature and relative humidity.

References

Chapter 2

1. Shields, J. *Adhesives Handbook, 3rd edn.*, Newnes-Butterworth, London, 1985.
2. Lark, R.J. The classification and control of epoxy adhesives in civil engineering, PhD Thesis, University of Dundee, November 1983.
3. Mika, T.F. Curing agents and modifiers, Chap. 4 in *Epoxy Resins – Chemistry and Technology* (Ed. May, C.A. and Tanaka, Y), 1973.
4. Lee, M. and Neville, K. *Handbook of Epoxy Resins*, McGraw Hill, 1967.
5. Garnish, E.W. Advances in epoxy adhesive technology, Chap. 3 in *Developments in Adhesives – 1* (Ed. Wake, W.C.), Applied Science Publishers, 1977.
6. Potter, W.G. *Epoxide Resins*, London Ilifree Books, 1970.
7. Lees, W.A. Toughened structural adhesives and their uses, *International Journal of Adhesion and Adhesives*, July 1981, pp. 241–7.
8. Wake, W.C. *Adhesion and the Formulation of Adhesives*, Applied Science Publishers, 1982.
9. Bolger. Structural adhesives for metal bonding, Chap. 1 in *Treatise on Adhesion and Adhesives, Vol. 3* (Ed. Patrick, R.L.), 1973.
10. Shaw, J.D.N. A review of resins used in construction, *International Journal of Adhesion and Adhesives*, V2, No. 2, April 1982, pp. 77–83.
11. Judge, A.I., Cheriton, L.W. and Lambe, R.W. Bonding systems for concrete repair – an assessment of commonly used materials, in *Adhesion Between Polymers and Concrete* (Ed. Sasse, H.R.), Chapman and Hall, 1986.
12. Department of Transport. *Materials for the Repair of Concrete Highway Structures*, Departmental Standard BD 27/86, 1986.
13. Department of Transport. *The Investigation and Repair of Concrete Highway Structures*, Departmental Advice Note BA 23/86, 1986.
14. MacKenzie, G.K. A study of freshly mixed epoxy resin adhesives for civil engineering, MSc thesis, University of Dundee, November 1986.
15. Federation Internationale de la Precontrainte. Proposal for a standard for acceptable tests and verification of epoxy bonding agents for segmental construction, Cement & Concrete Association, Slough, 1978.
16. British Standards Institution. BS 5350: Part B4, *British Standard Test Methods for Adhesives: Determination of Pot Life*, BSI, London, 1976.
17. British Standards Institution. DD88 Draft for Development, *Method*

References

of Assessment of Pot Life of Non-flowing Resin Compositions for use in Civil Engineering, BSI, London, 1983.

18. Packham, D.E. The adhesion of polymers to metals: the role of surface topography, in *Adhesion Aspects of Polymeric Coatings* (Ed. Mittall, K.L.), Plenum Press, New York, 1983, pp. 19–44.

19. Sika Ltd. *Sikadur 31 Adhesive Technical Notes.*

20. Permabond Adhesives Ltd. *Technical Data Permabond ESP 110.*

21. Mays, G.C. and Hutchinson, A.R. Engineering property requirements for structural adhesives, *Proceedings of the Institution of Civil Engineers*, Part 2, Vol 85., September 1988, pp. 485–501.

22. Hutchinson, A.R. and Lees, D.E. A compression test for determining adhesive material properties, *International Conference on Structural Adhesives in Engineering II* (Pub. Butterworth). Bristol University, Sept. 1989.

23. PERA. Adhesive bonding – a practical approach. PERA supplement to *Engineering Materials and Design*, 1987.

24. British Standards Institution. BS 2782 *Methods of Testing for Plastics*, BSI, London, 1970.

25. Emerson, M. Extreme values of bridge temperatures for design purposes, TRRL Report LR 744, Crowthorne, 1976.

26. Lark, R.J. and Mays, G.C. Epoxy adhesive formulation: its influence on civil engineering performance, in *Adhesion 9* (Ed. Allen, K.W.), Elsevier Applied Science Publishers, London, 1985, pp. 95–110.

27. Comyn, J. The relationship between joint durability and water diffusion, Chapter 8 in *Developments in Adhesives – 2* (Ed. Kinloch, A.J.), Applied Science Publishers, 1981.

28. Lloyd, J.O. and Calder, A.J. *The Microstructure of Epoxy Bonded Steel-to-Concrete Joints*, TRRL Supplementary Report 705, Crowthorne, 1982.

29. Albrecht, P., Mecklenburg, M.F. and Evans, B.M. *Screening of Structural Adhesives for Application to Steel Bridges*, Interim Report prepared for U.S. Dept. of Transportation, Federal Highway Administration, February 1985.

30. Dharmarajan, N., Kumar, S. and Armeniades, C.D. A constitutive equation for creep in polymer concretes and their resin binders, *The Production, Performance and Potential of Polymers in Concrete* (Ed. Staynes, B.W.), ICPIC '87, Brighton, Sept. 1987.

Chapter 3

1. Comyn, J. *Developments in Adhesives – 2* (Ed. Kinloch, A.J.), Applied Science Publishers, London, 1981, Chapter 8.

2. Kinloch, A.J. *Durability of Structural Adhesives*, Applied Science Publishers, London, 1983.

3. Brockmann, W. *International Conference on Structural Adhesives in Engineering*, Paper No. C176, Institution of Mechanical Engineers, Bristol University, July 1986.

4. Hutchinson, A.R. *Proceedings of International Conference on Structural Faults and Repair – 87*, (Ed. Forde, M.C.), London University, July 1987, p. 235.

References

5. Kinloch, A.J. (Ed.). *Developments in Adhesives – 2*, Applied Science Publishers, London, 1981.
6. Zisman, W.A. *Handbook of Adhesives, 2nd edn.*, Van Nostrand Rheinhold Company, New York, 1977, Chapter 3.
7. Kinloch, A.J. *Journal of Materials Science,* **15**, 1980, p. 2141.
8. Bateup, B.O. *International Journal of Adhesion and Adhesives,* **1**, No. 5, 1981, p. 233.
9. Huntsberger, J.R. *Journal of Adhesion,* **12**, 1981, p. 3.
10. Brewis, D.M. (Ed.). *Surface analysis and Pretreatment of Plastics and Metals*, Applied Science Publishers, London, 1982.
11. Wake, W.C. *Adhesion and the Formulation of Adhesives*, 2nd edn., Applied Science Publishers, London, 1982.
12. Mittal, K.L. *Polymer Eng. and Science,* **17**, No. 7, 1977, p. 467.
13. Fowkes, F.M. *Journal of Adhesion Science and Technology,* **1**, No. 1, 1987, p. 7.
14. Bolger, J.C. *Adhesion Aspects of Polymeric Coatings*, (Ed. Mittal, K.L.), Plenum Press, New York, 1983, p. 3.
15. Allen, K.W. *Adhesives in Engineering Design*, W.A. Lees, The Design Council, Springer-Verlag Publishers, London, 1984, Appendix 1.
16. Packham, D.E. *Adhesion Aspects of Polymeric Coatings*, Plenum Press, New York, 1983, p. 19.
17. Hewlett, P.C. and Pollard, C.A. *Adhesion – 1* (Ed. Allen, K.W.), Applied Science Publishers, London, 1977, Chapter 3.
18. Zisman, W.A. *Adhesion Science and Technology, Polymer Science and Technology, Vol. 9A* (Ed. Lee, L.H.), Plenum Press, New York, p. 55.
19. Gledhill, R.A., Kinloch, A.J. and Shaw, S.J. *Journal of Adhesion,* **9**, 1977, p. 81.
20. Gledhill, R.A. and Kinloch, A.J. *Journal of Adhesion,* **9**, 1974, p. 315.
21. Comyn, J. *Durability of Structural Adhesives* (Ed. Kinloch, A.J.), Applied Science Publishers, London, 1983, Chapter 3.
22. Bischof, C., Bauer, A., Kapelle, R. and Possart, W. *Int. J. Adhesion and Adhesives,* **5**, No. 2, 1985, p. 97.
23. Gettings, M. and Kinloch, A.J. *Journal of Materials Science,* **12**, 1977, p. 2511.
24. Gettings, M. and Kinloch, A.J. *Surface and Interface Analysis,* **1**, No. 5, 1979, p. 165 and **1**, No. 6, 1979, p. 189.
25. Snogren, R.C. *Handbook of Surface Preparation*, Communication Channels Pub., Atlanta GA, USA, 1974.
26. Kinloch, A.J. and Dukes, W.A. *National Seminar on Exploiting Adhesive Bonding in Production*, The Welding Institute, Rugby, December 1978, Paper 3.
27. Brewis, D.M. (Ed.). *Surface Analysis and Pretreatment of Plastics and Metals*, Applied Science Publishers, London, 1982.
28. Lees, W.A. *Adhesives in Engineering Design*, The Design Council, Springer-Verlag Pub., London, 1984.
29. Shields, J. Adhesives Handbook, 3rd edn., Newnes-Butterworth Pub., London, 1985.
30. CP3012, *Code of Practice for Cleaning and Preparation of Metal*

References

Surfaces, British Standards Institution, 1972.

31. BS5350, *Methods of Test for Adhesives: Adherend Preparation*, British Standards Institution, 1976, part A1.
32. ASTM D2651, *Preparation of Metal Surfaces for Adhesive Bonding*, American Society for Testing and Materials, Vol. 22, 1979.
33. FeRFA. *Application Guide NO. 2, 2nd edn.*, Federation of Resin Formulators and Applicators Ltd., October 1980.
34. Albericci, P. *Durability of Structural Adhesives* (Ed. Kinloch, A.J), Applied Science Publishers, London, 1983, p. 317.
35. Hine, P.J., Packham, D.E. and Mudarris, S. *The International Adhesion Conference, PRI*, Nottingham University, September 1984, Paper 10.
36. Beevers, A. and Njegic, A. *Conference on Adhesives, Sealants and Encapsulants*, London, November 1985, Poster paper, p. 349.
37. Allen, K.W., Greenwood, L. and Siwela, T.C. *Journal of Adhesion*, **16**, 1983, p. 127.
38. Bolger, J.C. *Treatise on Adhesion and Adhesives, Vol. 3* (Ed. Patrick, R.L.), Marcel Dekker Publishers, New York, 1973, Chapter 1.
39. Sykes, J.M. *Surface Analysis and Pretreatment of Plastics and Metals* (Ed. Brewis, D.M.), Applied Science Publishers, London, 1982, Chapter 7.
40. Haigh, I.P. (Ed.) *Painting Steelwork*, CIRIA Report 93, 1982.
41. Brockmann, W. *Durability of Structural Adhesives* (Ed. Kinloch, A.J.), Applied Science Publishers, London, 1983, p. 281.
42. Hutchinson, A.R. Durability of Structural Adhesive Joints, PhD Thesis, Dundee University, 1986.
43. BS4232, *Surface finish of blast-cleaned steel for painting*, British Standards Institution, 1967.
44. SIS 05 59 00, *Pictorial Surface Preparation Standards for Painting Steel Surfaces*, Swedish Standards Institution, 1967.
45. Adams, R.D. and Wake, W.C. *Structural Adhesive Joints in Engineering*, Elsevier Applied Science Publishers, London, 1984.
46. Trawinski, D.L. *Society for the Advancement of Materials and Process Engineering Quarterly*, October 1984, p. 1.
47. McIntyre, R.T. *Applied Polymer Symposia 19* (Ed. Bodnar, M.J.), Wiley & Sons Publishers, New York, 1972, p. 309.
48. Lees, W.A. *Adhesion 9* (Ed. Allen, K.W.), Elsevier Applied Science Publishers, London, 1985, Chapter 8.
49. Garnish, E.W. *Adhesion 2* (Ed. Allen, K.W.), Applied Science Publishers, London, 1978, Chapter 3.
50. Minford, J.D. *Adhesives Age*, July 1974, p. 24.
51. McMillan, J.C. *Bonded Joints and Preparation for Bonding*, NATO AGARD Lecture Series No. 102, March 1979, p. 7.1.
52. Venables, J.D., McNamara, D.K., Chen, J.M., Sun, T.S. and Hopping, R.L. *Applications of Surface Science 3*, North-Holland Publishers, 1979, p. 88.
53. Cotter, J.L. and Kohler, R. *International Journal of Adhesion and Adhesives*, **1**, No. 1, 1980, p. 23.
54. Brockmann, W., Hennemann, O.-D. and Kollek, H. *Int. J. Adhesion and Adhesives*, **2**, No. 1, 1982, p. 33.
55. Moloney, A.C. *Surface Analysis and Pretreatment of Plastics and*

References

Metals (Ed. Brewis, D.M.), Applied Science Pub., London, 1982, Chapter 8.

56. Venables, J.D. *Journal of Materials Science*, **19**, 1984, p. 1.
57. Kollek, H. *International Journal of Adhesion and Adhesives*, **5**, No. 2, 1985, p. 75.
58. Kinloch, A.J. *Journal of Adhesion*, **10**, 1979, p. 193.
59. Poole, P. and Watts, J.F. *International Journal of Adhesives*, **5**, No. 1, 1985, p. 33.
60. Hewlett, P.C. and Shaw, J.D.N. *Developments in Adhesives – 1* (Ed. Wake, W.C.), Applied Science Pub., London, 1977, Chapter 2.
61. Shaw, J.D.N. *International Journal of Adhesion and Adhesives*, **2**, No. 2, 1982, p. 77.
62. Sasse, H.R. and Fiebrich, M. Rilem: *Materials and Structures*, **16**, No. 94, July–August 1983, p. 293.
63. Concrete Society, Technical Report 26, 1984.
64. Gaul, R.W. *Concrete International*, **6**, No. 7, 1984, p. 17.
65. Allen, R.T.L. and Edwards, S.G. *The Repair of Concrete Structures*, Blackie & Son Publishers, London, 1987, p. 118.
66. Suprenant, B. and Malisch, W. *Concrete Construction*, November 1986, p. 927.
67. Hewlett, P.C. *Symposium on Protection of Concrete and Reinforcement*, Paint Research Association/TRRL, Cement and Concrete Association, Slough, May 1987, Paper 1.
68. Murray, A. McC. and Long, A.E. *2nd International Conference on Structural Faults and Repair*, I.C.E., London, Pub. Engineering Technics Press, 1985, p. 323.
69. BS6319. *Testing of Resin Compositions for Use in Construction*, British Standards Institution, 1983.
70. Bowditch, M.R. and Stannard, K.J. *Adhesives, Sealants and Encapsulants Conference* (ASE 85), London, Day 3, 5–7 November 1985, p. 66.
71. Hugenschmidt, F. *International Journal of Adhesion and Adhesives*, **2**, No. 2, 1982, p. 84.
72. Marsden, J.G. and Sterman, S. *Handbook of Adhesives, 2nd edn.* (Ed. Skeist, I), Van Nostrand Reinhold Company, New York, 1977, Chapter 40.
73. Rosen, M.R. *Journal of Coatings Technology*, **50**, No. 644, 1976, p. 70.
74. Plueddemann, E.P. *Silane Coupling Agents*, 1st edn., Plenum Press, New York, 1982.
75. Monte, S.J., Sugerman, G. and Spindel, S. 71st Annual Meeting of the AIChE on *The Construction Industry's Response to Changing Resource and Materials Economics*, Miami Beach, Florida, November 1978.
76. Monte, S.J. and Sugerman, G. SPE NATEC Paper, October 1982.
77. KEN-REACT Reference Manual. *Titanate Coupling Agents for Filled Polymers*, Kenrich Petrochemicals Inc., New Jersey, U.S.A. (U.K. Sales: Hubron Sales Ltd., 2A Partington Street, Failsworth, Manchester M35 0RD).
78. Comyn, J. *Course on Design and Assembly with Engineering Adhesives*, Cranfield Institute of Technology, May 1984.

References

References

79. Allen, K.W. and Stevens, M.G. *Journal of Adhesion*, **14**, 1982, p. 137.
80. Walker, P. *Journal of Oil and Colour Chemists Association*, **65**, 1982, p. 415; **65**, 1982, p. 436; **66**, 1983, p. 188; **67**, 1984, p. 108; **67**, 1984, p. 126.
81. Pfiefer, D.W. and Scali, M.J. National Cooperative Highway Research Programme Report No. 244, U.S. Transportation Research Board, Washington D.C., December 1981.
82. Middelboe, S. *New Civil Engineer*, 9 May 1985, p. 14.
83. Wessex Resins and Adhesives, 189–193 Spring Road, Sholing, Southampton SO2 7NY.
84. Burns, A.L. *Adhesives, Sealants and Encapsulants Conference* (ASE 85), London, Day 3, 5–7 November 1985, p. 79.
85. Graham, L. and Emerson, J.A. *Adhesion Aspects of Polymeric Coatings* (Ed. Mittal, K.L.), Plenum Press, New York, p. 395.
86. ISO 4624, *Paints and Varnishes – Pull-off Test for Adhesion*, International Organisation for Standardisation, First edn., 1978–07–01.
87. Sickfeld, J. *Adhesion Aspects of Polymeric Coatings* (Ed. Mittal, K.L), Plenum Press, New York, 1983, p. 543.
88. Marceau, J.A., Moji, Y. and McMillan, J.C. *Adhesives Age*, October 1977, p. 28.
89. Davis, G.D. and Venables, J.D. *Durability of Structural Adhesives* (Ed. Kinloch, A.J.), Applied Science Publishers, London, 1983, Chapter 2.
90. Brewis, D.M., Comyn, J., Oxley, D.P., Pritchard, R.G., Reynolds, S., Werrett, C.R. and Kinloch, A.J. *Surface and Interface Analysis*, **6**, 1984, p. 40.
91. Comyn, J., Kinloch, A.J., Horley, C.C., Mallik, R.R., Oxley, D.P., Pritchard, R.G., Reynolds, S. and Werrett, C.R. *International Journal of Adhesion and Adhesives*, **5**, No. 2, 1985, p. 59.

Chapter 4

1. Shields, J. *Engineering Design Guide 02*, Oxford University Press, 1974.
2. Skeist, I. (Ed.). *Handbook of Adhesives*, Von Nostrand Reinhold Co., New York, 1977.
3. Kinloch, A.J. (Ed.). *Developments in Adhesives – 2*, Applied Science Publishers, London, 1981.
4. Kinloch, A.J. *Journal of Materials Science*, **17**, 1982, p. 617.
5. Adams, R.D. and Wake, W.C. *Structural Adhesive Joints in Engineering*, Elsevier Applied Science Publishers, London, 1984.
6. Lees, W.A. *Adhesives in Engineering Design*, The Design Council, Springer-Verlag Pub., London, 1984.
7. Shields, J. *Adhesives Handbook, 3rd edn.*, Newnes-Butterworths Pub., London, 1985.
8. Kinloch, A.J. *Journal of Materials Science*, **15**, 1980, p. 2141.
9. Petronio, M. *Handbook of Adhesives, 2nd edn.* (Ed. Skeist, I), Van Nostrand Reinhold Co., New York, 1977, Chapter 6.

References

10. BS 5350. *Methods of Test for Adhesives*, British Standards Institution, London, 1976, part C5.
11. ASTM D1002. Strength Properties of Adhesives in Shear by Tension Loading (Metal-to-Metal), *Annual Book of Standards*, **15.06**, American Society for Testing and Materials, 1986.
12. Gordon, J.E. *Structures – or why things don't fall down*, Penguin Books, London, 1978.
13. de Bruyne, N.A. *Aircraft Engineering*, Workshop and Production Section, April 1944, p. 115.
14. Mylonas, C. and de Bruyne, N.A. *Adhesion and Adhesives*, (Ed. de Bruyne, N.A. and Houwink, R.), Elsevier Publishers, Amsterdam, 1951, p. 91.
15. Volkersen, P. *Lüftfahrtforschung*, **15**, 1938, p. 41.
16. Goland, M. and Reissner, E. *Journal of Applied Mechanics*, Trans. ASME, **66**, 1944, A17–A27.
17. Renton, W.J. and Vinson, J.R. *Journal of Adhesion*, **7**, No. 3, 1975, p. 175.
18. Allman, D.J. *Mechanics and Applied Mathematics*, **30**, 1977, part 4, p. 415.
19. Hart-Smith, L.J. *Developments in Adhesives – 2* (Ed. Kinloch, A.J.), Applied Science Publishers, London, 1981, Chapter 1.
20. ESDU 80039. Engineering Sciences Data Unit, London, December 1980.
21. ESDU 79016. Engineering Sciences Data Unit, London, September 1979.
22. Armstrong, K.B. *International Journal of Adhesion and Adhesives*, **3**, No. 1, 1983, p. 37.
23. Harris, J.A. and Adams, R.D. *International Journal of Adhesion and Adhesives*, **4**, No. 2, 1984, p. 65.
24. Griffith, A.A. *Philosophical Transactions of the Royal Society* A221, 1921, p. 163.
25. Anderson, G.P., Bennett, S.J. and De Vries, K.L. *Analysis and Testing of Adhesive Bonds*, Academic Press Inc., London, 1977.
26. Bascom, W.D., Cottington, R.L. and Timmons, C.O. Durability of Adhesive Bonded Structures (Ed. Bodnar, M.J.), *Journal of Applied Polymer Science*, Applied Polymer Symposia 32, J. Wiley and Sons, New York, 1977, p. 165.
27. Bascom, W.D. and Hunston, D.L. *Adhesion 6* (Ed. Allen, K.W.), Applied Science Publishers, London, 1982, Chapter 14.
28. Kinloch, A.J. and Shaw, S.J. *Developments in Adhesives – 2* (Ed. Kinloch, A.J.), Applied Science Publishers, London, 1981, p. 83.
29. Ripling, E.J., Mostovoy, S. and Patrick R.L. *Applications of Fracture Mechanics to Adhesive Joints*, ASTM STP 360, 1963, p. 5.
30. Irwin, G.R. *Applied Materials Research*, **3**, 1964, p. 65.
31. Parker, A.P. *The Mechanics of Fracture and Fatigue*, E. and F.N. Spon Ltd., London, 1981.
32. Renton, W.J. U.S. Govt. Report AFML-TR-78-127, **1**: AD-A065 500; **2**: AD-A065 521, September 1978.
33. Romanko, J. and Knauss, W.G. U.S. Govt. Report AFWAL-TR-80 4037, **1**: AD-A087 280; **2**: AD-A087 360, 1980.
34. Albrecht, P., Mecklenburg, M.F. and Evans, B.M. Interim Report

References

FHWA/RD, Federal Highway Administration, Contract DTFH61-84-R-0027, Washington DC, U.S.A., February 1985.
35. Albericci, P. *Durability of Structural Adhesives* (Ed. Kinloch, A.J.), Applied Science Publishers, London, 1973, p. 317.
36. Fairbairn, W. *Philosophical Transactions of the Royal Society*, London, 1850, part 2, p. 677.
37. Wake, W.C. *Adhesion and the Formulation of Adhesives, 2nd edn.*, Applied Science Publishers, London, 1982.
38. Lutz, P. *Adhesives, Sealants and Encapsulants Conference* (ASE 85), Kensington, London, 5–7 November 1985, Day 2, p. 34.
39. Weissberg, V. and Arcan, M. ASTM STP 981 (Ed. Johnson, W.W), American Society for Testing and Materials, Philadelphia, 1988, pp. 28–38.
40. Krieger, M.R.B. *International Conference on Structural Adhesives in Engineering*, Institute of Mechanical Engineers, Bristol University, July 1986, Paper C174.
41. Althof, W.J., Klinger, G., Newmann, G. and Schlothauer, J. Royal Aircraft Establishment, Library Translation 1999, Farnborough U.K., January 1979.
42. Althof, W. *International Adhesion Conference*, Nottingham University, September 1984, Poster paper 18.
43. ESDU 81022. Engineering Sciences Data Unit, London, July 1981.
44. de Bruyne, N.A. *Adhesion and Adhesives* (Ed. de Bruyne, N.A. and Houwink, R.), Elsevier Publishers, Amsterdam, 1951, p. 91.
45. Sage, G.N. and Adams, R.D. ESDU Memo 42, Engineering Sciences Data Unit, London, June 1982.
46. Bryant, R.W. and Dukes, W.A. Structural Adhesives Bonding (Ed. Bodnar, M.J.), *Journal of Applied Polymer Science*, Applied Polymer Symposia 3, 1966, p. 81.
47. Stringer, L.G. *Journal of Adhesion*, **18**, No. 3, 1985, p. 185.
48. Lin, C.J. and Bell, J.P. *Journal of Applied Polymer Science*, **16**, 1972, p. 1721.
49. ASTM E229-70(81). Shear Strength and Shear Modulus of Structural Adhesives, *Annual Book of Standards*, **15.06**, American Society for Testing and Materials, 1986.
50. BS 6319, *Testing of Resin Compositions for Use in Construction*, British Standards Institution, London, 1984, Part 4.
51. Eyre, J.R. and Domone, P.L.J. *2nd International Conference on Structural Faults and Repair*, I.C.E., London University, Engineering Technics Press, 1985, p. 141.
52. ASTM D897-78(83). Tensile properties of Adhesive Bonds, *Annual Book of Standards*, **15.06**, American Society for Testing and Materials, 1986.
53. Brockmann, W. *Durability of Structural Adhesives* (Ed. Kinloch, A.J.), Applied Science Publishers, London, 1983, p. 281.
54. ASTM D903-49(83). Peel or Stripping Strength of Adhesive Bonds, *Annual Book of Standards*, **15.06**, American Society for Testing and Materials, 1986.
55. ASTM D1876-72(83). Peel Resistance of Adhesives (T-Peel Test), *Annual Book of Standards*, **15.06**, American Society for Testing and Materials, 1986.

References

56. BS 5350. *Methods of Test for Adhesives*, British Standards Institution, London, 1987–1980, Parts C9-C14.
57. Marceau, J.A. and Scardino, W.M. U.S. Govt. Report AFML-TRL73-3, Amer. Acc. No. AD-A008 528, February 1975.
58. Marceau, J.A., Moji, Y. and McMillan, J.C. *Adhesives Age*, October 1977, p. 28.
59. McMillan, J.C. Bonded Joints and preparation for Bonding, *NATO AGARD Lecture Series No. 102*, March 1979, paper 7–1.
60. McMillan, J.C. *Developments in Adhesives – 2* (Ed. Kinloch, A.J.), Applied Science Publishers, London, 1981, Chapter 7.
61. Thrall, E.W. *Adhesion 4* (Ed. Allen, K.W.), Applied Science Publishers, London, 1980, Chapter 1.
62. Arnold, D.B. *Developments in Adhesives – 2* (Ed. Kinloch, A.J.), Applied Science Publishers, London, 1981, Chapter 6.
63. Stone, M.H. and Peet, T. Royal Aircraft Establishment Tech. Memo MAT 349, Farnborough U.K., July 1980.
64. Krieger, R.B. *5th National Technical Conference of the Society for the Advancement of Materials and Process Engineering (SAMPE)*, October 1973, p. 643.
65. ASTM D3762-79 (83). Adhesive-bonded Surface Durability of Aluminium (Wedge Test), *Annual Book of Standards,* **15.06**, American Society for Testing and Materials, 1986.
66. Allen, K.W., Chan, S.Y.T. and Armstrong, K.B. *International Journal of Adhesion and Adhesives*, **2**, No. 4, 1982, p. 239.
67. Allen, K.W. Hatzinikolaou, T. and Armstrong, K.B. *International Journal of Adhesion and Adhesives*, Vol 4, No. 3, 1984, p. 133.
68. Brewis, D.M. *Durability of Structural Adhesives* (Ed. Kinloch, A.J.), Applied Science Publishers, London, 1983, Chapter 5.
69. Kinloch, A.J. (Ed.). *Durability of Structural Adhesives*, Applied Science Publishers, London, 1983.
70. Bascom, W.D. , Timmons, C.O. and Jones, R.L. *Journal of Materials Science,* **10**, 1975, p. 1037.
71. Allen, K.W. and Shanahan, M.E.R. *Journal of Adhesion,* **7**, 1975, p. 161.
72. Allen, K.W. and Shanahan, M.E.R. *Journal of Adhesion,* **8**, 1976, p. 43.
73. Althof, W. and Brockmann, W. *Adhesives Age*, September 1977, p. 27.
74. Romanko, J. and Knauss, W.G. *Developments in Adhesives – 2* (Ed. Kinloch, A.J.), Applied Science Publishers, London, 1981, Chapter 5.
75. Ro nanko, J., Liechti, K. and Knauss, W.G. U.S. Govt. Report AFWAL-TR-82-4139, Amer. Acc. No. AD-A124 324, 1982.
76. Romanko, J. Bonded Joints and Preparation for Bonding, *NATO AGARD Lecture Series No. 102*, March 1979, paper 4–1.
77. Krieger, R.B. *Adhesives Age*, June 1978, p. 26.
78. Krieger, R.B. *24th National SAMPE Symposium and Exhibition*, San Francisco, U.S.A., May 1979.
79. Marceau, J.A., McMillan, J.C. and Scardino, W.M. *Adhesives Age,* **21**, No. 4, 1977, p. 37.
80. Matting, A. and Draugelates, U. *Adhesion,* **11**, 1968, No. 1, p. 5;

No. 3, p. 110; No. 4, p. 161.

81. Allen, K.W., Smith, S.M., Wake, W.C. and van Raalte, A.O. *International Journal of Adhesion and Adhesives*, **5**, No. 1, 1985, p. 23.

82. Mays, G.C. and Harvey, W.J. *IABSE Colloquium on Fatigue of Steel and Concrete Structures*, paper 37, Lausanne, Switzerland, 1982, p. 393.

83. Mays, G.C. and Tilly, G.P. *International Journal of Adhesion and Adhesives*, **2**, No. 2, 1982, p. 109.

84. Kinloch, A.J. and Young, R.J. *Fracture Behaviour of Polymers*, Applied Science Publishers, London, 1983.

85. Ferry, J.D. *Viscoelastic Properties of Polymers*, J. Wiley & Sons, New York, 1970.

86. Hunston, D.L., Carter, W.T. and Rushford, J.L. *Developments in Adhesives – 2* (Ed. Kinloch, A.J.), Applied Science Publishers, London, 1981, Chapter 4.

87. Lark, R.J. and Mays, G.C. *Adhesion 9* (Ed. Allen, K.W.), Elsevier Applied Science Publishers, London, 1985, Chapter 7.

88. Hunston, D.L. and Bullmann, G.W. *International Journal of Adhesion and Adhesives*, **5**, No. 2, 1985, p. 69.

89. Kinloch, A.J. *Journal of Adhesion*, **10**, 1979, p. 193.

90. Comyn, J. *Developments in Adhesives – 2* (Ed. Kinloch, A.J.), Applied Science Publishers, London, 1981, Chapter 8.

91. Comyn, J. *Durability of Structural Adhesives* (Ed. Kinloch, A.J.), Applied Science Publishers, London, 1983, Chapter 3.

92. Hockney, M.G.D. Royal Aircraft Establishment, Farnborough U.K., Technical Report 73016, Trial 1, Part 2, January 1973; Trial 2, Part 2, October 1973.

93. Cotter, J.L. *Developments in Adhesives – 1* (Ed. Wake, W.C.), Applied Science Publishers, London, 1977, Chapter 1.

94. Brewis, D.M., Comyn, J. and Shalash, R.J.A. *International Journal of Adhesion and Adhesives*, **2**, No. 4, 1982, p. 215.

95. Althof, W. *Adhesion 5* (Ed. Allen, K.W.), Applied Science Publishers, London, 1981, Chapter 2.

96. Sargent, J.P. and Ashbee, K.H.G. *Journal of Physics D: Applied Physics*, **14**, 1981, p. 1933.

97. Moncrieff, A. and Weaver, G. *Cleaning, Science for Conservators Book 2*, The Crafts Council, London, 1983.

98. Adamson, M.J. *Journal of Materials Science*, **15**, 1980, p. 1736.

99. Cairns, D.S. and Adams, D.F. *Journal of Reinforced Plastics and Composites*, 1985.

100. Sargent, J.P. and Ashbee, K.H.G. *Journal of Adhesion*, **11**, 1980, p. 175.

101. Butt, R.I. and Cotter, J.L. *Journal of Adhesion*, **8**, 1976, p. 11.

102. Antoon, M.K. and Koenig, J.L. *Journal of Macromolecular Science – Reviews of Macromolecular Chemistry*, C19(1), 1980, p. 135.

103. Hutchinson, A.R. Durability of Structural Adhesive Joints, PhD Thesis, Dundee University, 1986.

104. Brockmann, W. *Aspects of Polymeric Coatings* (Ed. Mittal, K.L.), Plenum Press, New York, 1983, p. 265.

105. Brockmann, W. *International Conference on Structural Adhesives in*

Engineering, Institute of Mechanical Engineers, Bristol University, July 1986, Paper C176.

106. Gledhill, R.A. and Kinloch, A.J. *Journal of Adhesion,* **9**, 1974, p. 315.
107. Sykes, J.M. *Surface Analysis and Pretreatment of Plastics and Metals* (Ed. Brewis, D.M.), Applied Science Publishers, London, 1982, Chapter 7.
108. Bascom, W.D. *Adhesives Age,* April 1979, p. 28.
109. Cherry, B.W. and Thomson, K.W. *Adhesion 1* (Ed. Allen, K.W.), Applied Science Publishers, London, 1977, Chapter 16.
110. Cherry, B.W. and Thomson, K.W. *Adhesion 4* (Ed. Allen, K.W.), Applied Science Publishers, London, 1980, Chapter 5.
111. Bolger, J.C. *Adhesion Aspects of Polymeric Coatings*, (Ed. Mittal, K.L.), Plenum Press, New York, 1983, p. 3.
112. Mastronardi, P., Carfagna, C. and Nicholais, L. *Journal of Materials Science,* **18**, 1983, p. 1977.
113. Bowditch, M.R. and Stannard, K.J. *Adhesives, Sealants and Encapsulants Conference* (ASE 85), Kensington, London, 5–7 November 1985, Day 3, p. 66.
114. Gettings, M. and Kinloch, A.J. *Journal of Materials Science,* **12**, 1977, p. 2511.
115. Hutchinson, A.R. *Proceedings of the International Conference on Structural Faults and Repair – 87* (Ed. Forde, M.C.), London University, July 1987, p. 235.
116. Krieger, R.B. Private Communication, 1984.
117. Andrews, E.H. and Stevenson, A. *Journal of Adhesion,* **11**, 1980, p. 17.
118. Bethune, A.W. *Journal of the Society for the Advancement of Materials and Process Engineering Quarterly,* **11**, No. 4, 1975, p. 4.
119. Bodnar, M.J. (Ed.). Durability of Adhesive Bonded Structures, *Journal of Applied Polymer Science*, Applied Polymer Symposia 32, J. Wiley & Sons, New York, 1977.
120. Brewis, D.M., Comyn, J. and Tegg, J.L. *International Journal of Adhesion and Adhesives,* **1**, No. 1, 1980, p. 35.
121. Brockmann, W. *Adhesives Age,* July 1974, p. 24.
122. Minford, J.D. *International Journal of Adhesion and Adhesives,* **2**, No. 1, 1982, p. 25.
123. Minford, J.D. *Durability of Structural Adhesives* (Ed. Kinloch, A.J.), Applied Science Publishers, London, 1983, Chapter 4.
124. Venables, J.D. *Journal of Materials Science,* **19**, 1984, p. 1.
125. Blight, G.E. Concrete Beton, No. 20, Part 12, 1980, p. 7.
126. Calder, A.J.J. Transport and Road Research Laboratory, Crowthorne U.K., SR 529, 1979; Internal Note 0330/80, June 1980.
127. Lloyd, G.O. and Calder, A.J.J. Transport and Road Research Laboratory, Crowthorne U.K., SR 705, 1982.
128. Garnish E.W. *Adhesion 2* (Ed. Allen, K.W.), Applied Science Publishers, London, 1978, Chapter 3.
129. Gledhill, R.A., Kinloch, A.J. and Shaw, S.J. *Journal of Adhesion,* **11**, 1980, p. 3.
130. Gettings, M. and Kinloch, A.J. *Surface and Interface Analysis,* **1**, No. 5, 1979, p. 165; **1**, No. 6, 1979, p. 189.

References

131. Jones, R. and Swamy, R.N. *2nd International Conference on Structural Faults and Repair*, I.C.E., London University, Engineering Technics Press, 1985.
132. Ladner, M. and Weder, C. Concrete structures with bonded external reinforcement. EMPA Report No. 206, Dubendorf, 1981.
133. Nara, H. and Gasparini, D., U.S. Department of Transportation, Federal Highway Administration Report FHWA/OH-81/011, Case Western Reserve University, Cleveland Ohio, U.S.A., 1981.
134. Stevenson, A. *International Journal of Adhesion and Adhesives*, **5**, No. 2, 1985, p. 81.
135. Trawinski, D.L. *Journal of the Society for the Advancement of Materials and Process Engineering Quarterly*, October 1984, p. 1.
136. Walker, P. *Journal of Oil and Colour Chemists' Association*, **65**, 1982, p. 415; **65**, 1982, p. 436; **66**, 1983, p. 188; **67**, 1984, p. 108; **67**, 1984, p. 126.

Chapter 5

1. Mays, G.C. and Hutchinson, A.R. *Proceedings of the Institution of Civil Engineers, Part 2*, **85**, pp. 485–501.
2. Less, W.A. *Adhesives in Engineering Design*, The Design Council, Springer-Verlag Pub., London, 1984.
3. Lees, W.A. (Ed.). *Adhesives and the Engineer*, Permabond Adhesives Limited, Mech. Eng. Pub., London, 1989.
4. Shields, J. *Adhesives Handbook, 3rd Edition*, Newnes-Butterworths Pub., London, 1985.
5. Adams, R.D. and Wake, W.C. *Structural Adhesive Joints in Engineering*, Elsevier Applied Science Pub., London, 1984.
6. Schliekelmann, R.J. *International Conference on Structural Adhesives in Engineering*, Bristol University, Proceedings of the Institution of Mechanical Engineers, 1986, p. 241.
7. BS 5350. *Methods of Test for Adhesives*, British Standards Institution, London, 1978–80.
8. BS 6319. *Testing of Resin Compositions for use in Construction*, British Standards Institution, London 1983–87.
9. FIP/9/2. Proposal for a standard for acceptance tests and verification of epoxy bonding agents for segmented construction, Federation Internationale de la Precontrainte, 1978.
10. Schliekelmann, R.J. *NATO AGARD Lecture Series*, **102**, 1979, 8–1.
11. Segal, E. and Rose, J.L. *Research Techniques in Non-Destructive Testing, IV*, Academic Press, London, 1980, p. 275.
12. Stone, D.E.W. *2nd International Conference on Adhesives, Sealants and Encapsulants (ASE 86)*, London, 4–6 November 1986.
13. Guyott, C.C.H., Cawley, P. and Adams, R.D. *Journal of Adhesion*, **20**, No. 2, 1986, pp. 129–59.
14. Adams, R.D. and Cawley, P. *NDT International*, **21**, 1988, pp. 208–22.
15. Mackie, R.J. and Vardy, A.E. *2nd International Conference on Structural Adhesives in Engineering*, Bristol University, 20–22 September 1987, Butterworth Scientific Pub., pp. 211–15.

References

Chapter 6

1. Concrete Society. Non-structural cracks in concrete, Technical Report No. 22, Concrete Society, London, 1982.
2. Concrete Society. Repair of concrete damaged by reinforcement corrosion, Technical Report No. 26, Concrete Society, London, 1984.
3. Allen, R.T.L. and Edwards, S.C. Repairs to cracked concrete. Chap. 5 in *Repair of Concrete Structures* (Ed. Allen & Edwards), Blackie, Glasgow, 1987.
4. BS 882: 1983. *Aggregates from Natural Sources for Concrete*, British Standards Institution, London.
5. Klopper, M. *Bautenschutz Bausenurung*, 1, No. 3, 1987, pp. 86–96.
6. Minkarah, I. and Ringo, B.C. Behaviour and repair of deteriorated reinforced concrete beams, Transportation Research Record 821. University of Cincinnatti, 1982.
7. El-Jazairi, B. The requirements of hardened MPC mortars and concrete relevant to the requirements of rapid repair of concrete pavements, *Concrete* 21, No. 9, September 1987, pp. 25–32.
8. Emberson, N.K. and Mays G.C. Polymer mortars for the repair of structural concrete: the significance of property mismatch, *The Production, Performance and Potential of Polymers in Concrete* (Ed. Staynes, B.W.), Fifth International Congress on Polymers in Concrete, ICPIC '87, Brighton, September 1987, pp. 335–42.
9. ACI Committee 546. Guide for repair of concrete bridge superstructures, Report No. ACI 546. IR-80, *Concrete International*, September 1980, pp. 69–88.
10. Cusens, A.R. and Smith D.W. *The Structural Engineer, Part A,* 58, 1980, pp. 13–18.
11. Hinterwaldner, R. *Adhäsion*, 1977, 21, pp. 13.
12. BS 6319: Part 4: 1984. *Testing of resin compositions for use in construction*, Part 4, Method for measurement of bond strength (slant shear method), British Standards Institution, London.
13. Dixon, J.F. and Sunley, V.K. Use of bond coats in concrete repair, *Concrete,* 17, No. 8, August 1983, pp. 34–5.
14. Tabor, L.J. Twixt old and new: achieving a bond when casting fresh concrete against hardened concrete, *Proceedings of the 2nd International Conference on Structural Faults and Repair, 1985,* Engineering Technics Press, Edinburgh, pp. 57–63.
15. Raithby, K.D. External strengthening of concrete bridges with bonded steel plates, Transport and Road Research Laboratory Supplementary Report 612, Crowthorne 1980.
16. Ladner, M. and Weder, C. Concrete structures with bonded external reinforcement, EMPA Report No. 206, Dubendorf, 1981.
17. Lark, R.J. and Mays G.C. Epoxy adhesive formulation: its influence on civil engineering performance. In *Adhesion 9* (Ed. Allen, K.W.), Applied Science Publishers, 1984.
18. Mays G.C. and Hutchinson, A.R. Engineering property requirements for structural adhesives, *Proceedings of the Institution of Civil Engineers, Part 2,* 85, September 1988, pp. 485–501.
19. Scottish Development Agency. Adhesive bonding in a civil engineering

environment, A market evaluation of certain adhesive bonding techniques, December 1983.

20. Swedish Standards Institution. *Pictorial Surface Preparation Standards for Painting Steel Surfaces*, SIS 055900, Stockholm, 1967.

21. Long, A.E. Montgomery, F.R. and Cleland, D. Assessment of concrete strength and durability on site, *Proceedings of the International Conference on Structural Faults and Repair* (Ed. Forde, M.C.), London, July 1987, pp. 61–73.

22. Mander, R.F. Use of resins in road and bridge construction and repair, Seminar on resins in construction, The Federation of Epoxy Resin Formulators and Applicators, London, October 1979.

23. Lees, W.A. Toughened structural adhesives and their uses, *International Journal of Adhesion and Adhesives*, July 1981, pp. 241–47.

24. Nara, H. and Gasparini, D. Fatigue resistance of adhesively bonded connections, Report No. FHWA/OH-81/011, Case Western Reserve University, Cleveland, Ohio, November 1981.

25. BS 5400: Part 10: 1980. Steel, concrete and composite bridges, Part 10, Code of Practice for Fatigue, British Standards Institution, London.

26. Mays, G.C. *Fatigue and Creep Performance of Epoxy Resin Adhesive Joints*, To be published in TRRL Contractors Report Series, 1990.

27. Standing Committee on Structural Safety. 5th Report 1982, *The Structural Engineer* **61A**, No. 2, February 1983.

28. Dembleton, B. Stuck fast, *New Civil Engineer*, 16 March 1989, pp. 36–7.

29. Hashim, S.A., Cowling, M.J. and Winkle, I.E. Design and assessment methodologies for adhesively bonded structural connections, *Structural Adhesives in Engineering II '89*, University of Bristol, September 1989.

30. Bowden, C. Restoring life to old timbers, *Building Trades Journal*, **187**, No. 5540, 2 February 1984.

31. Wurtzburg, S. A challenging facade refurbishment, *Construction Repair*, **3**, No. 3, March 1989.

32. Danby, J. Preserving integrity and appearance, *Construction Repairs and Maintenance*, **2**, No. 2, March 1986.

Chapter 7

1. Mander, R.F. Use of resins in road and bridge construction and repair. *International Journal of Cement Composites*, **3**, Pt. 1, February 1981, pp. 27–39.

2. Hewlett, P.C. and Shaw, J.D.N. Structural adhesives used in civil engineering, in *Developments in Adhesives* (Ed. Wake, W.C.), Applied Science Publishers, London, 1977.

3. James, J.G. Calcined bauxite and other artificial polish resistant road stones, TRRL Technical Paper No. LR 84, Crowthorne, 1967.

4. Durst, W.J. The Shellgrip surfacing system, *Highways and Road Construction*, **42**, 1974, pp. 12–16.

5. Paterson, W.S. and Ravenhill, K. Reinforcement connector and anchorage methods, CIRIA Report 92, CIRIA, London, 1981.

References

6. Mayfield, B., Bates, M.W. and Snell, C. *Civil Engineering*, March 1978, p. 63.
7. Sims, F.A. The use of resins in bridges and structural engineering in West Yorkshire, Seminar on resins in construction, The Federation of Epoxy Resin Formulators and Applicators, London, October 1979.
8. Anon. German bonded bridge, *Construction Industry International*, **5**, No. 4, April 1979, pp. 31–7.
9. Lerchenthal, H. and Rosenthal, J. Flexural behaviour of concrete slabs reinforced with steel sheet. *Materiaux et Constructions*, **15**, No. 88, pp. 279–82.
10. Mays, G.C. and Vardy, A.E. Adhesive bonded steel-concrete composite construction, *International Journal of Adhesion & Adhesives*, **2**, No 2, April 1982, pp. 103–7.
11. Anon. Advanced technology on M180, *Concrete*, **12**, No. 10, October 1978, pp. 16–20.
12. Barfoot, J. Steady progress on Cardiff's 'alternative tender' viaduct, *Concrete*, **18**, No. 4, April 1984, pp. 11–14.
13. Middleboe, S. Welsh bridge failure starts tendon scare, *New Civil Engineer*, 12 December 1985, p. 6.
14. Federation Internationale de la Precontrainte. Proposal for a standard for acceptance tests and verification of epoxy bonding agents for segmental construction, FIP/9/2, March 1978.
15. Arvid Grant & Associates Inc. Segments of cable stayed bridge adhesive bonded, Adhesives/Coatings/Sealants, *Construction*, **13**, No. 1, May 1978.
16. Moreton, A.J. Epoxy glue joints in precast concrete segmental bridge construction, *Proceedings of the Institution of Civil Engineers, Part 1*, No. 70, February 1981, pp. 163–77.
17. Clifton, J.R., Beaghley, H.F. and Mathey R.G. Non-metallic coatings for concrete reinforcing bars, Coating materials, US National Bureau of Standards, Technical Note 768, August 1975.
18. Safier, A.S. Developments and use of electrostatic epoxy-powder coated reinforcement, *The Structural Engineer*, **67**, No. 6, March 1989, pp. 95–8.
19. ASTM: A775-81. *Standard Specification for Epoxy-coated Steel Reinforcing Bars*, American Society for Testing Materials.
20. Anon. Epoxy coating earns share of European market, *Concrete*, **22**, No. 11, November 1988, pp. 14–15.
21. Bishop, R.R. The problems of specifying coated rebars in the UK, Symposium on Protection of Concrete Reinforcement, Paint Research Association, May 1987.
22. Treadaway K.W.J. and Davies, H. Performance of fusion-bonded epoxy-coated steel reinforcement. *The Structural Engineer*, **67**, No. 21, March 1989, pp. 99–108.
23. Tilley, G.P. Durability of concrete bridges, *Journal of the Institution of Highways and Transportation*, **35**, No. 2, February 1988, pp. 10–19.
24. BS 5268: Part 2: 1984. *Code of Practice for the Structural Use of Timber*, Part 2, Permissible stress design, materials and workmanship, British Standards Institution, London.
25. BS 1204: Part 1: 1979. *Synthetic resin adhesives (phenolic and aminoplastic) for wood*, Part 1, Specification for gap-filliing adhesives,

References

British Standards Institution, London.

26. TRADA. Introduction to the specification of glued laminated members, Wood Information Sheet Section 1 Sheet 5, Timber Research and Development Association, September 1985.

Chapter 8

1. Smith, D.W. and Cusens, A.R. *Symposium on Main Trends in the Development of Steel Structures and Modern Methods for their Fabrication*, IABSE, Moscow, September 1978, pp. 47–53.
2. Mays, G.C. and Vardy, A.E. *International Journal of Adhesion and Adhesives*, **2**, No. 2, 1982, pp. 103–7.
3. Mays, G.C. *et al. International Journal of Cement Composites and Lightweight Concrete*, **5**, No. 3, 1983, pp. 151–63.
4. British Standards Institution. *Steel, Concrete and Composite Bridges*, BS 5400, 1978.
5. Wong, C.K. and Vardy, A.E. *The Structural Engineer*, **63B**, No. 1, 1985, pp. 8–13.
6. Kam, J.B. MSc Thesis, University of Dundee, 1985.
7. Wong, C.K. Finite Prism Analysis in Open Sandwich Construction, PhD Thesis, University of Dundee, 1982.
8. British Standards Institution. *Code of Practice for the Structural Use of Concrete*, CP110, 1972.
9. Ong, K.C.G. Open Sandwich Construction for Bridge Decks, PhD Thesis, University of Dundee, 1981.
10. Dakers, J.J. The Design and Construction of Wester Duntanlich Bridge Deck using the Open Sandwich Slab Technique, MSc Thesis, University of Dundee, 1985.
11. *The New Civil Engineer*, 30 June 1988, p. 31.
12. Anon. Construction Industry International, Vol. 5, No. 4, 1979, pp. 31–37.
13. Hutchinson, A.R. and Beevers, A. BRITE EURAM 2nd Tech. Days, Brussels, Jan–Feb 1989, pp. 129–33.
14. Harvey, W.J. and Vardy, A.E. Bonded Stiffeners for Steel Bridges, *Proceedings of the 27th Conference on Adhesion and Adhesives*, City University, London, March 1989.
15. Hashim, S.A., Cowling, M.J. and Winkle, I.E. *International Conference on Structural Adhesives in Engineering II*, University of Bristol, Butterworth Scientific, September 1989, pp. 10–17.
16. Albrecht, P. *et al. Application of Adhesives to Steel Bridges*, Report No. FHWA/RD-84/073, Federal Highway Administration, McLean, Va., USA, 1984.
17. Albrecht, P. and Sahli, A.H. *ASTM Special Technical Publication 927*, American Society for Testing and Materials, Philadelphia, Pa, USA, 1985, pp. 72–94.
18. Albrecht, P. *Journal of Structural Engineering, ASCE*, **113**, No. 6, June 1987, pp. 1236–50.
19. Soetens, F. Adhesive Joints in Adluminium Alloy Structures, TNO-IBBC Report No. B-89-665, Delft, The Netherlands, September 1989.
20. Soetens, F. *International Conference on Structural Adhesives in*

References

Engineering II, University of Bristol, Butterworth Scientific, September 1989, pp. 18–27.

21. Navaratnarajah, V. *Indian Journal of Adhesion and Adhesives*, **3**, No. 2, 1983, pp. 93–9.
22. Metal-Bond Technology Limited. 61 Bankhall Lane, Hale, Altrincham, Cheshire, England.
23. Holloway, L.C. (Ed.). *Polymers and Polymer Composites in Construction*, Thomas Telford Limited, London, 1990.
24. Anon. *The Structural Engineer*, **67**, No. 12, 1989, p. A8.
25. Extren fibreglass structural members from the Morrison Molded Fiber Glass (MMFG) Company. For UK sales, contact Fibreforce Composites Limited, Fairbank Lane, Whitehouse, Runcorn, Cheshire, England.
26. *The New Civil Engineer*, 18 August 1988, p. 14.
27. Saadatmanesh, H. and Ehsani, M.R. *Concrete International, Design & Construction*, **12**, No. 3, 1990, pp. 65–71.
28. *The New Civil Engineer*, 14 April 1983, p. 22.
29. *The New Civil Engineer*, 11 July 1985, p. 36.
30. Retrofix. The Pitchmastic Building Group, Royds Works, Attercliffe Road, Sheffield.

Appendix

1. Mays, G.C. and Hutchinson, A.R. Engineering Property Requirements for Structural Adhesives, *Proceedings of the Institution of Civil Engineers, Pt 2*, **85**, September 1988, pp. 485–501.
2. Mackenzie, G.K. A study of freshly mixed epoxy resin adhesives for civil engineering, University of Dundee, 1986, MSc thesis.
3. Hutchinson, A.R. Durability of structural adhesive joints, University of Dundee, 1986, PhD thesis.
4. Lark, R.J. and Mays, G.C. Epoxy adhesive formulation: its influence on civil engineering performance, *Adhesion 9* (Ed. Allen, K. W.). Elsevier Applied Science Publishers, 1985, pp. 95–110.
5. British Standards Institution. *Methods of Testing Plastics*, BSI, London, 1976, BS 2782: many parts.
6. Swedish Standards Institution. *Pictorial Surface Preparation Standards for Painting Steel Surfaces*, SIS, Stockholm, 1967, SIS 055900.
7. Marceau, J.A. *et al.* A wedge test for evaluating adhesive-bonded surface durability, *Adhesives Age*, October 1977, pp. 28–34.
8. Hutchinson, A.R. Surface pretreatment – the key to durability, *Conference on Structure Faults and Repairs – 87*, London University, July 1987.
9. British Standards Institution. *Methods of Test for Adhesives*, BSI, London, 1976, BS 4350: Parts A–H.
10. Mays, G.C. Fatigue performance and durability of structural adhesive joints, University of Dundee, 1985, PhD thesis.

Index

326

iron, 140
steel, 138
brittle fracture, 120, 288
buckling tests, 283, 288
buildings, 12–14, 218–20, 236, 269,
    281, 290
buoys, 20
bush hammering, 103
butt joint tests, 59, 140, 146, 147

cables, 253, 291
calcined bauxite, 245
capillary action, 66, 82, 165
car bodies, 23
carbonation, 203–4
carbon dioxide, 53, 200, 204, 226
carbon fibre reinforced plastic
    (CFRP), 11, 13, 20, 104, 170, 293
casein, 18, 19
catalyst, 42, 204
cellulose, 30
cement, 47
ceramic tiles, 12
chemical deterioration, 66
chemical pretreatment, 92, 94–7, 98,
    103
chemical resistance, 42, 92
chemisorption, 169, 170, 172
chipboard, 7
chlorides, 93, 203–4, 260
chlorinated rubber, 204
civil engineering, 14–19, 25–8, 180,
    227, 231, 267, 279, 290, 297
cladding
    attachments, 291–2
    panels, 12,280
cleanliness, 111
cleavage
    strength, 122
    tests, 140
climatic exposure, 172–3
closed form solutions, 125–9
closed sandwich construction, 277–8
coal tar, 171
coatings, 9, 16, 106, 203–5
    powder, 20
cohesion, 163, 168
    control, 190–3
coin tap test, 192
cold formed sections, 279, 288
collagen, 30
column repair, 214–15
compliance, 69, 73, 152, 153
    spectrum, 183, 297–308
composite construction, 18, 20, 215,
    254–5, 269, 271, 279–80, 297

composite materials, 8, 10, 11, 13, 19,
    20, 22, 45, 104, 117, 170, 183, 290
composite pretreatment, 104–5
composite pultrusions, 281
compound structural elements, 280–1
compression test, 56
concrete, 9, 117, 147, 160, 170, 183,
    186
    overlays, 215–17, 219
    patches, 199–203
    precast, 12, 21, 219, 254, 256, 279
    pretreatment, 101–4
    repair, 45, 197–215, 290, 297
    strengthening, 215–31, 297
    substrate, 84
condensation, 103, 111, 187
construction joint, 223
contact angle, 53, 79, 80–2, 113, 190
contamination, 77, 84, 88, 94, 103,
    105, 111, 120, 216, 229
correlation diagram, 125
corrosion, 125, 163, 167, 169, 170,
    173, 177, 252, 260
    protection, 20, 105, 272
cost, 90, 100, 185, 197, 217, 256, 277,
    280, 293
counter casting, 25
couplers, 253, 290
coupling agents, 42, 86, 101, 105–11,
    170, 171, 298
cover plates, 223, 285–7
cracks
    adhesive, 66, 74, 86, 91, 130–2,
        149, 152, 165, 225
    concrete, 197–9, 215, 218, 220, 229,
        260
    mortar, 243
crane rails, 18
crazing, 40, 66, 165
creep, 8, 33, 43, 56, 67–74, 117, 122,
    127, 138, 157–8, 183, 211–14, 226,
    227, 232, 249, 260, 271, 287, 290,
    301
critical surface tension, 84
cure, 8, 54, 185, 200, 225, 229–30,
    271
    pressure, 10, 12
    rate, 42, 202, 243
    temperature, 10, 12, 27, 32, 44,
        226, 227, 230, 259, 263, 271, 297,
        299
    time, 299
curling, 243
curtain walling, 12, 280, 288, 289
cyanoacrylate, 32
cycloaliphatic amines, 37–8